TI-83 Graphing Calculator Guide

for

Moore and McCabe's

Introduction to the Practice of Statistics
Fifth Edition

David K. Neal
Western Kentucky University

W. H. Freeman and Company
New York

ISBN-13: 978-0-7167-6364-2
ISBN-10: 0-7167-6364-8

Printed in the United States of America

Second printing

W. H. Freeman and Company
41 Madison Avenue
New York, NY 10010
Houndmills, Basingstoke RG21 6XS, England
www.whfreeman.com

Preface

The study of statistics has become commonplace in a variety of disciplines and the practice of statistics is no longer limited to specially trained statisticians. The work of agriculturists, biologists, economists, psychologists, sociologists, and many others now quite often relies on the proper use of statistical methods. However, it is probably safe to say that most practitioners have neither the time nor the inclination to perform the long, tedious calculations that are often necessary in statistical inference. Fortunately there are now software packages and calculators that can perform many of these calculations in an instant, thus freeing the user to spend valuable time on methods and conclusions rather than on computation.

With its built-in statistical features, the TI-83 Plus Graphing Calculator has revolutionized the teaching of statistics. Students and teachers now have instant access to many statistical procedures. Advanced techniques can be programmed into the TI-83 Plus which then make it as powerful as, but much more convenient than, common statistical software packages.

This manual serves as a companion to *Introduction to the Practice of Statistics* (5th Edition) by David S. Moore and George P. McCabe. Problems from each section of the text are worked using either the built-in TI-83 Plus functions or programs specially written for this calculator. The tremendous capabilities and usefulness of the TI-83 Plus are demonstrated throughout. It is hoped that students, teachers, and practitioners of statistics will continue to make use of these capabilities, and that readers will find this manual to be helpful.

Programs

All codes and instructions for the programs are provided in the manual; however, they can be downloaded directly from **http://www.wku.edu/~david.neal/ips5e/**

Acknowledgments

I would like to thank all those who have used the first two editions of this manual. I appreciate the feedback that I have received from students and instructors who have found these instructions to be helpful. My thanks go to W. H. Freeman and Company for giving me another opportunity to revise the manual for the 5th edition of *IPS*. Special thanks go to editorial assistant Sarah Fleischman for her organization and help in keeping me on schedule. As always, my sincere gratitude goes to Professors Moore and McCabe for providing educators and students with an excellent text for studying the practice of statistics.

David K. Neal
Department of Mathematics
Western Kentucky University
Bowling Green, KY 42101

email: david.neal@wku.edu
homepage: http://www.wku.edu/~david.neal/

Contents

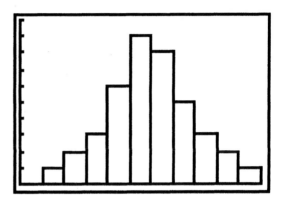

Looking at Data— Distributions

Introduction

In this chapter, we use the TI-83 Plus to view data sets. We first show how to make bar graphs, histograms, and time plots. Then we use the calculator to compute basic statistics, such as the mean, median, and standard deviation, and show how to view data further with boxplots. Lastly, we use the TI-83 Plus for calculations involving normal distributions.

1.1 Displaying Distributions with Graphs

We start by using the TI-83 Plus to graph data sets. In this section, we will use the **STAT EDIT** screen to enter data into lists and use the **STAT PLOT** menu to create bar charts, histograms, and time plots.

Throughout the manual, we will be working with data that is entered into lists **L1** through **L6** on the TI-83 Plus. These lists can be found in the **STAT EDIT** screen. A list should be cleared before entering new data into it.

Press STAT, then 1
to bring up the list editor.

Use cursor to highlight a list,
press CLEAR, then press
ENTER to clear the list.

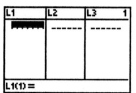

Enter new data
into a cleared list.

Bar Graph of Categorical Data

Exercise 1.14 Here are the percents of women among students seeking various graduate and professional degrees during the 1999–2000 academic year.

Degree	Percent female
MBA	39.8
MAE	76.2
Other MA	59.6
Other MS	53.0
Ed.D.	70.8
Other Ph.D.	54.2
MD	44.0
Law	50.2
Theology	20.2

Make a bar graph of the data. Make another bar graph with the bars ordered by height.

Solution. First, label the nine categories as 1–9 and enter these values into list **L1**, then enter the percents into list **L2**. Next, adjust the **WINDOW** and **STAT PLOT** settings and graph. Here we use X from 1 to 10 on a scale of 1 in order to see all nine bars, and use Y from 0 to 100 to represent the range 0% to 100%.

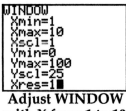

Enter 1–9 into list
L1 and enter the
percents into list L2.

Adjust WINDOW
with X from 1 to 10
and Y from 0 to 100.

Press STAT PLOT,
then 1. Set to the
third type to plot L1
with frequencies L2.

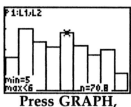

Press GRAPH,
then TRACE.

Next, we make a *Pareto* chart with the bars ordered by height. To do so, we simply use the **SortD(** command from the **STAT EDIT** screen to sort the data in list **L2** into descending order, then re-graph.

 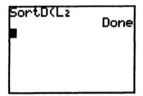 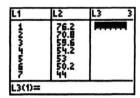 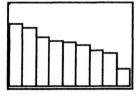

Making a Histogram

Exercise 1.34 Make a histogram of Cavendish's measurements of the density of the earth.

5.50	5.61	4.88	5.07	5.26	5.55	5.36	5.29	5.58	5.65
5.57	5.53	5.62	5.29	5.44	5.34	5.79	5.10	5.27	5.39
5.42	5.47	5.63	5.34	5.46	5.30	5.75	5.68	5.85	

Solution. First, we enter the data into a list. Here we will use list **L3**. We will graph using an X range of 4.8 to 6 on a scale of 0.1 in the **WINDOW** settings. As with a bar graph, we use the third type in the **STAT PLOT** settings for a histogram. Next, we set the **Xlist** to L3 with frequencies of 1, and press **GRAPH**.

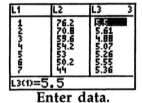   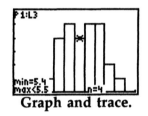

Enter data.   Adjust WINDOW.   Adjust STAT PLOT.   Graph and trace.

Time plot

The next exercise demonstrates how to use a time plot to view data observations that are made over a period of time.

Exercise 1.28 Here are data on the recruitment (in millions) of new fish to the rock sole population in the Bering Sea between 1973 and 2000. Make a time plot of the recruitment.

Year	Recruitment	Year	Recruitment	Year	Recruitment	Year	Recruitment
1973	173	1980	1411	1987	4700	1994	505
1974	234	1981	1431	1988	1702	1995	304
1975	616	1982	1250	1989	1119	1996	425
1976	344	1983	2246	1990	2407	1997	214
1977	515	1984	1793	1991	1049	1998	385
1978	576	1985	1793	1992	505	1999	445
1979	727	1986	2809	1993	998	2000	676

Solution. Enter the data into lists and adjust the **WINDOW** so that X ranges through the years and Y ranges through the recruitment values. Adjust the **STAT PLOT** settings to the second type and set the lists to those that contain the data, then graph.

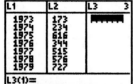  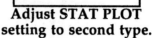 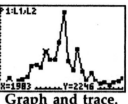

| Enter data. | Adjust WINDOW. | Adjust STAT PLOT setting to second type. | Graph and trace. |

Note: We also can use the **seq(** command from the **LIST OPS** menu to enter the years into a list. On the **Home** screen, simply enter the command **seq(K,K,1973,2000)➔L1**.

1.2 Describing Distributions with Numbers

In this section, we will use the **1–Var Stats** command from the **STAT CALC** menu to compute the various statistics of a data set including the mean, standard deviation, and five-number summary. We also will use boxplots and modified boxplots to view these statistics.

1–Var Stats

Exercise 1.74 Find $\bar{x}$ and s for Cavendish's data from Exercise 1.34. Also give the five-number summary and create a boxplot to view the spread.

5.50	5.61	4.88	5.07	5.26	5.55	5.36	5.29	5.58	5.65
5.57	5.53	5.62	5.29	5.44	5.34	5.79	5.10	5.27	5.39
5.42	5.47	5.63	5.34	5.46	5.30	5.75	5.68	5.85	

Solution. After the data has been entered into a list, say list **L1**, we compute the statistics by entering the command **1–Var Stats L1**. The values of $\bar{x}$ and s are then displayed. Scroll down to see the five-number summary.

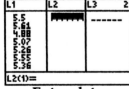 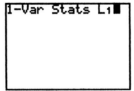 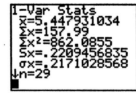 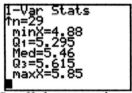

| Enter data. | Compute 1-Var Stats. | View $\bar{x}$ and s. | Scroll down to view five-number summary. |

Two standard deviation values are given. The first, **Sx**, is the standard deviation that is denoted by s in the text. Thus, we obtain $\bar{x} \approx 5.448$ and $s \approx 0.221$ with a five-number summary of $4.88 - 5.295 - 5.46 - 5.615 - 5.85$.

Boxplot

To make a boxplot of data in a list, first adjust the **WINDOW** settings so that **Xmin** to **Xmax** includes the entire range of the measurements. (The boxplot ignores the Y range). Next, adjust the **STAT PLOT** settings to the fifth type for a boxplot, denote the list that contains the data, and press **ENTER**. Press **TRACE** and scroll to see the values of the five-number summary.

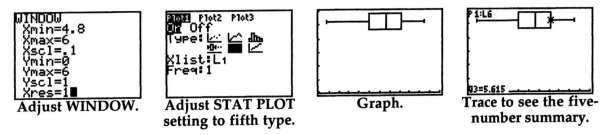

| Adjust WINDOW. | Adjust STAT PLOT setting to fifth type. | Graph. | Trace to see the five-number summary. |

Exercise 1.52 The data from Exercise 1.35 is given below. The values give the nightly study time claimed by samples of first-year college men and women. (a) Compute $\bar{x}$ and s for these data sets. (b) For each data set, find the suspected outliers as determined by the $1.5 \times IQR$ rule. (c) Make side-by-side modified boxplots of the two data sets.

Women					Men				
180	120	180	360	240	90	120	30	90	200
120	180	120	240	170	90	45	30	120	75
150	120	180	180	150	150	120	60	240	300
200	150	180	150	180	240	60	120	60	30
120	60	120	180	180	30	230	120	95	150
90	240	180	115	120	0	200	120	120	180

Solution. We enter the data sets into separate lists and then use the **1–Var Stats** command on each list. We then adjust the window settings so that the X range includes the span of both sets.

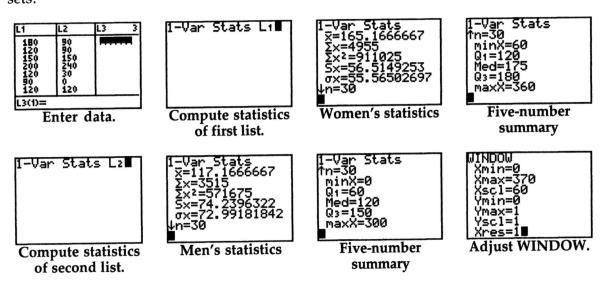

| Enter data. | Compute statistics of first list. | Women's statistics | Five-number summary |
| Compute statistics of second list. | Men's statistics | Five-number summary | Adjust WINDOW. |

Note: If we have two data sets with an equal number of measurements, then we can compute the statistics of both simultaneously with the **2–Var Stats** command from the **STAT CALC** menu. In this case, enter **2–Var Stats L1,L2**. However, this command does not display the five-number summaries.

From the five-number summaries, we can compute boundaries according to the $1.5 \times IQR$ rule. In each case, we need the values $Q_1 - 1.5 \times (Q_3 - Q_1)$ and $Q_3 + 1.5 \times (Q_3 - Q_1)$. For the women's study times, these values are

$$120 - 1.5 \times (180 - 120) = \mathbf{30} \quad \text{and} \quad 180 + 1.5 \times (180 - 120) = \mathbf{270}$$

For the men's study times, these values are

$$60 - 1.5 \times (150 - 60) = \mathbf{-75} \quad \text{and} \quad 150 + 1.5 \times (150 - 60) = \mathbf{285}$$

Now we can determine the suspected outliers. For the women, these outliers are any times below 30 minutes or above 270 minutes, while for the men they are any times below $^-$75 minutes or above 285 minutes. To see these values more quickly, we can use the **SortA(** command from the **STAT EDIT** menu to sort each list into increasing order. Enter **SortA(L1** then **SortA(L2**. Because these lists have the same size, we can also enter the command **SortA(L1,L2**. In each case, there are no low outliers, but the time 360 is a high outlier for the women and the time 300 is a high outlier for the men.

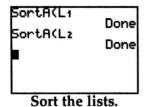

Sort the lists.

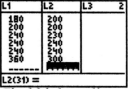

Find low outliers.

Find high outliers.

Next, we will make side-by-side boxplots of both data sets followed by modified side-by-side boxplots that denote the single outlier for each.

Adjust Plot1 settings to the fifth type for list L1.

Adjust Plot2 settings to the fifth type for list L2.

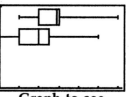

Graph to see side-by-side boxplots.

Adjust Plot1 settings to the fourth type for list L1.

Adjust Plot2 settings to the fourth type for list L2.

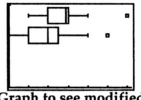

Graph to see modified side-by-side boxplots.

1.3 Density Curves and Normal Distributions

The TI-83 Plus has several commands in the **DISTR** menu that can be used for graphing normal distributions, computing normal probabilities, and making inverse normal calculations. In this section, we demonstrate these various functions.

 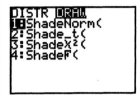

Plotting and Shading a Normal Distribution

Example 1.23 The distribution of heights of young women are approximately normal with mean μ = 64.5 inches and standard deviation σ = 2.5 inches. (a) Plot a density curve for this $N(64.5, 2.5)$ distribution. (b) Shade the region and compute the probability of heights that are within one standard deviation of average.

Solution. (a) We must enter the normal density function **normalpdf(X, μ, σ)** and adjust the window settings before graphing.

Bring up the Y= screen. Go to the DISTR screen and enter 1 for the normalpdf(function. Type the line normalpdf(X,64.5,2.5) in Y1.

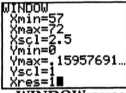

In the WINDOW screen, set Xmin to 64.5 − 3×2.5 and Xmax to 64.5 + 3×2.5. Set Ymin to 0 and set Ymax to the value 1 / √(2π) /2.5.

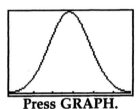

Press GRAPH.

(b) Return to the **Home** screen and bring up the command **ShadeNorm(** from the **DISTR DRAW** menu. Then Type and enter the command **ShadeNorm(64.5 − 2.5, 64.5 + 2.5, 64.5, 2.5)**, or in general, **ShadeNorm(lower, upper, μ, σ)**.

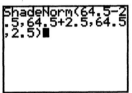   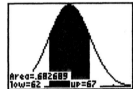

Exercise 1.93 Let $Z \sim N(0, 1)$ be the standard normal distribution. Shade the areas and find the proportions for the regions (a) $Z > 1.67$, (b) $-2 < Z < 1.67$.

Solution. (a) For the standard normal density function, we set the **WINDOW** with X ranging from −3 to 3 and Y ranging from 0 to $1/\sqrt{(2\pi)}$. Here we use **1E99** as an estimate for the upper bound of $+\infty$ in the **ShadeNorm(** command.

Adjust WINDOW.

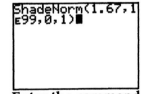

Enter the command ShadeNorm(1.67,1E99,0,1).

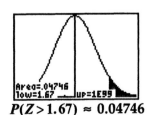

$P(Z > 1.67) \approx 0.04746$

(b) Before drawing a new graph, enter the command **ClrDraw** from the **CATALOG** in order to clear the shading from the previous graph. We find that $P(-2 < Z < 1.67) \approx 0.92979$.

 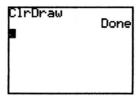 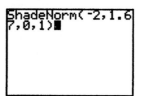 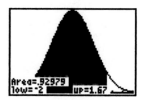

The Normal Distribution and Inverse Normal Commands

For a $N(\mu, \sigma)$ distribution X, we can also find probabilities with the built-in **normalcdf(** command from the **DISTR** menu. The command is used as follows:

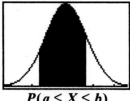 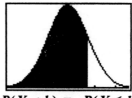 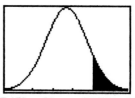

$P(a \leq X \leq b)$ $P(X < k) = P(X \leq k)$ $P(X > k) = P(X \geq k)$

normalcdf(a, b, μ, σ)   normalcdf(*1E 99, k, μ, σ)   normalcdf(k, 1E 99, μ, σ)

To find the value x for which $P(X \leq x)$ equals a desired proportion p (an inverse normal calculation), we use the command **invNorm(p, μ, σ)**. The following exercises demonstrate these commands.

Exercise 1.115 The lengths of human pregnancies are approximately normally distributed with a mean of 266 days and a standard deviation of 16 days. (a) What percent of pregnancies last fewer than 240 days? (b) What percent of pregnancies last between 240 and 270 days? (c) How long do the longest 20% of pregnancies last?

Solution. For parts (a) and (b), we simply use the **normalcdf(** command for $X \sim N(266, 16)$ by entering **normalcdf(*1E 99, 240, 266, 16)** and **normalcdf(240, 270, 266, 16)**. For part (c), we must find x such that $P(X \geq x) = 0.20$, which is equivalent to $P(X \leq x) = 0.80$. Thus, we enter the command **invNorm(.80, 266, 16)**.

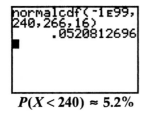 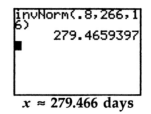

$P(X < 240) \approx 5.2\%$ $P(240 < X < 270) \approx 54.66\%$ $x \approx 279.466$ days

Exercise 1.97 The Weschler Adult Intelligence Scale (WAIS) provides IQ scores that are normally distributed with a mean of 100 and a standard deviation of 15. (a) What percent of adults would score 130 or higher? (b) What scores contain the middle 80% of all scores?

Solution. (a) We let $X \sim N(100, 15)$ and enter **normalcdf(130, 1E 99, 100, 15)**. (b) If 80% of scores are between x and y, then 10% of scores are below x and 10% of scores are above y. So x is the inverse normal of 0.10 and y is the inverse normal of 0.90.

$P(X \geq 130) \approx 2.275\%$

$80.777 < X < 119.223$
Contains 80% of the scores.

Normal Quantile Plot

Exercise 1.123 Make a normal quantile plot of Cavendish's data from Exercise 1.34.

5.50	5.61	4.88	5.07	5.26	5.55	5.36	5.29	5.58	5.65
5.57	5.53	5.62	5.29	5.44	5.34	5.79	5.10	5.27	5.39
5.42	5.47	5.63	5.34	5.46	5.30	5.75	5.68	5.85	

Solution. To see a normal quantile plot, we adjust the **WINDOW** so that **X** represents the data and so that **Y** ranges from –3 to 3, which covers most of the standard normal curve. In the **STAT PLOT** screen, the sixth type is the normal quantile plot. For Cavendish's data, we observe that the plot is nearly a straight line, with only a couple of outliers.

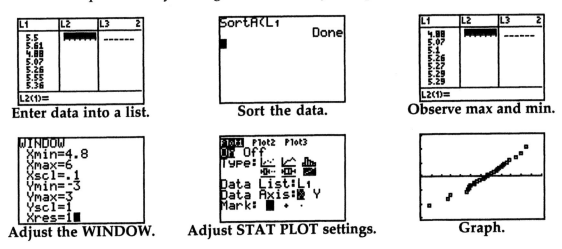

Enter data into a list.   Sort the data.   Observe max and min.

Adjust the WINDOW.   Adjust STAT PLOT settings.   Graph.

Exercise 1.126 Generate 100 observations from the standard normal distribution. Make a histogram of these observations. Make a normal quantile plot of the data.

Solution. We can generate a random list from a normal distribution with the **randNorm(** command from the **MATH PRB** menu. In general, to store n random values from a $N(\mu, \sigma)$ distribution into list **L1**, enter the command **randNorm(μ, σ, n)⟶L1**. Then adjust the **WINDOW** and **STAT PLOT** settings and graph to see a histogram.

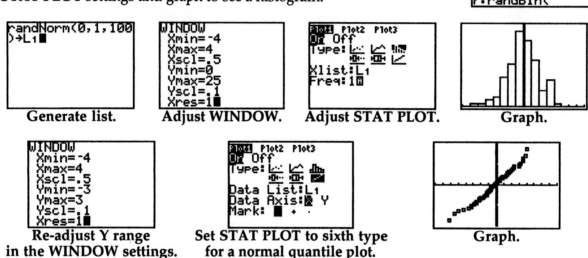

Generate list.   Adjust WINDOW.   Adjust STAT PLOT.   Graph.

Re-adjust Y range in the WINDOW settings.   Set STAT PLOT to sixth type for a normal quantile plot.   Graph.

Note: We can also generate data for measurements that are *uniformly* distributed from 0 to b. In this case, we use the **rand** command from the **MATH PRB** menu, and enter **b*rand(n)⟶L1** to store n of these values into list **L1**. (See Exercise 1.127 in the text.)

CHAPTER

2

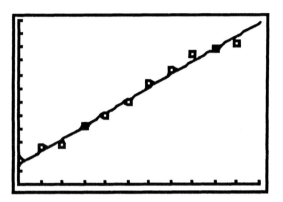

Looking at Data— Relationships

2.1 Scatterplots
2.2 Correlation
2.3 Least-Squares Regression
2.4 Cautions about Correlation and Regression

Introduction

In this chapter, we use the TI-83 Plus to graph the relationship between two quantitative variables using a scatterplot. We then show how to compute the correlation and find the least-squares regression line through the data. Lastly, we show how to work with the residuals of the regression line.

11

2.1 Scatterplots

We begin by showing how to make a scatterplot of two quantitative variables along the x and y axes so that we may observe if there is any noticeable relationship. In particular, we look for the strength of the linear relationship.

Exercise 2.9 Make a scatterplot of brain activity level against social distress score.

Subject	Social distress	Brain activity	Subject	Social distress	Brain activity
1	1.26	−0.055	8	2.18	0.025
2	1.85	−0.040	9	2.58	0.027
3	1.10	−0.026	10	2.75	0.033
4	2.50	−0.017	11	2.75	0.064
5	2.17	−0.017	12	3.33	0.077
6	2.67	0.017	13	3.65	0.124
7	2.01	0.021			

Solution. We first enter the data into the **STAT EDIT** screen. Here we use **L1** for the social distress scores, plotted on the x axis, and use **L2** for the brain activity levels that will be plotted on the y axis. We adjust the **WINDOW** as below so that the ranges include all measurements, and adjust the **STAT PLOT** settings by highlighting and entering the first **Type** and setting the appropriate lists. Press **GRAPH** and then press **TRACE** if so desired.

Enter data into lists.   Adjust WINDOW.

Adjust STAT PLOT.

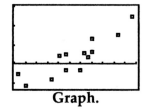

Graph.

We see that as the social distress score increases, then the brain activity level generally tends to increase also.

Exercise 2.13 Make a scatterplot of metabolic rate versus body mass for the females. Make another scatterplot with a different symbol for the males, and then combine the two plots.

Sex	Mass	Rate	Sex	Mass	Rate
M	62.0	1792	F	40.3	1189
M	62.9	1666	F	33.1	913
F	36.1	995	M	51.9	1460
F	54.6	1425	F	42.4	1124
F	48.5	1396	F	34.5	1052
F	42.0	1418	F	51.1	1347
M	47.4	1362	F	41.2	1204
F	50.6	1502	M	51.9	1867
F	42.0	1256	M	46.9	1439
M	48.7	1614			

Solution. We first enter the mass and rate of just the females into lists **L3** and **L4** respectively. Then we enter the mass and rate of the males into lists **L5** and **L6**. However, we adjust the **WINDOW** so that the X range includes all the masses and the Y range includes all the rates. We adjust the **STAT PLOT** settings in **Plot1** to obtain the scatterplot of **L3** versus **L4**, and adjust the **STAT PLOT** settings in **Plot2** to obtain the scatterplot of **L5** versus **L6**.

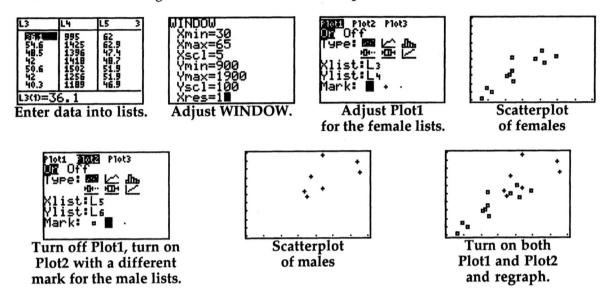

Enter data into lists.   Adjust WINDOW.   Adjust Plot1 for the female lists.   Scatterplot of females

Turn off Plot1, turn on Plot2 with a different mark for the male lists.   Scatterplot of males   Turn on both Plot1 and Plot2 and regraph.

Exercise 2.19 Make a plot of the total return against market sector. Compute the mean return for each sector, add the means to the plot, and connect the means with line segments.

Market sector	Fund returns (percent)						
Consumer	23.9	14.1	41.8	43.9	31.1		
Financial services	32.3	36.5	30.6	36.9	27.5		
Technology	26.1	62.7	68.1	71.9	57.0	35.0	59.4
Natural resources	22.9	7.6	32.1	28.7	29.5	19.1	

Solution. We will plot the market sectors on the x axis as the values 1, 2, 3, and 4. Because there are multiple returns for each sector, we enter each of the values 1 through 4 as many times into list **L1** as there are returns for that sector. We enter the corresponding returns into list **L2**.

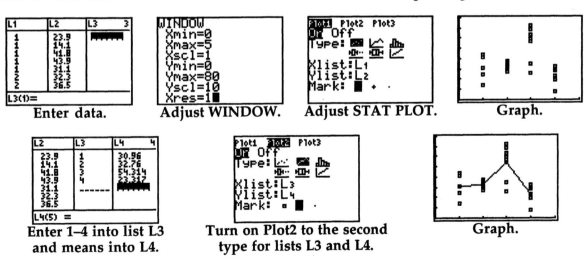

Enter data.   Adjust WINDOW.   Adjust STAT PLOT.   Graph.

Enter 1–4 into list L3 and means into L4.   Turn on Plot2 to the second type for lists L3 and L4.   Graph.

2.2 Correlation

In this section, we use the TI-83 Plus to compute the correlation coefficient r between paired data of quantitative variables.

First, we must make sure that the calculator's diagnostics are turned on. Enter the **CATALOG** (**2nd 0**) and scroll down to the **DiagnosticOn** command. Press **ENTER** to bring the command to the **Home** screen, then press **ENTER** again.

Exercise 2.29 The table below gives the heights in inches for a sample of women and the last men whom they dated. (a) Make a scatterplot. (b) Compute the correlation coefficient r between the heights of these men and women. (c) How would r change if all the men were 6 inches shorter than the heights given in the table?

Women (x)	66	64	66	65	70	65
Men (y)	72	68	70	68	71	65

Solution. (a) We enter the heights of the women into list **L1** and the heights of the men into list **L2**, adjust the **WINDOW** and **STAT PLOT** settings, and graph.

 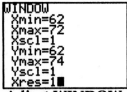  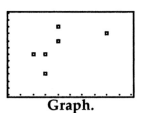

| Enter data. | Adjust WINDOW. | Adjust STAT PLOT. | Graph. |

(b) To compute the correlation, we use the **LinReg(a+bx)** command (item 8) from the **STAT CALC** menu. Enter the command **LinReg(a+bx) L1,L2**. The **LinReg(ax+b)** command (item 4) will also compute the correlation.

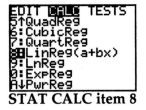 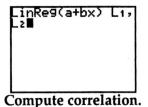 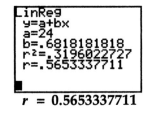

| STAT CALC item 8 | Compute correlation. | $r = 0.5653337711$ |

(c) We enter heights for all the males that are 6 inches shorter into list **L3**. Then we use the command **LinReg(a+bx) L1,L3** to see that r has not changed.

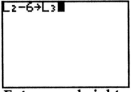  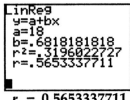

| Enter new heights. | View new heights. | Compute correlation. | $r = 0.5653337711$ |

2.3 Least-Squares Regression

In this section, we will compute the least-squares line of two quantitative variables and graph it through the scatterplot of the variables. We will also use the line to predict the y-value that should occur for a given x-value.

Exercise 2.47 The data from Exercise 2.9 are given below. (a) What is the equation of the least-squares regression line for predicting brain activity from social distress score? Make a scatterplot with this line drawn on it. (b) Use the equation of the regression line to get the predicted brain activity level for a distress score of 2. (c) What percent of the variation in brain activity among these subjects is explained by the straight-line relationship with social distress score?

Subject	Social distress	Brain activity	Subject	Social distress	Brain activity
1	1.26	−0.055	8	2.18	0.025
2	1.85	−0.040	9	2.58	0.027
3	1.10	−0.026	10	2.75	0.033
4	2.50	−0.017	11	2.75	0.064
5	2.17	−0.017	12	3.33	0.077
6	2.67	0.017	13	3.65	0.124
7	2.01	0.021			

Solution. (a) We obtain the linear regression line using the same **LinReg(a+bx)** command that computes the correlation. After entering data into lists, say **L1** and **L2**, enter the command **LinReg(a+bx) L1,L2**.

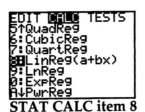  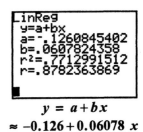

Enter data into lists. STAT CALC item 8 Compute the regression line.

$y = a + bx$
$\approx -0.126 + 0.06078\,x$

To graph the regression line, we must enter it into the **Y=** screen. We can type it in directly, or we can access this regression function from the **VARS Statistics EQ** menu. After entering the equation of the line into **Y1**, adjust the **WINDOW** and **STAT PLOT** settings and graph.

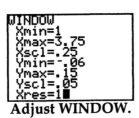 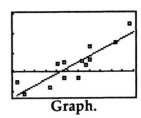

Go to Y=, press VARS, then 5, scroll right to EQ, press 1 to enter the line. Adjust WINDOW. Adjust STAT PLOT. Graph.

(b) Once the equation of the regression line is entered into the **Y=** screen, we can use the **CALC** screen to evaluate the line for a specific x value.

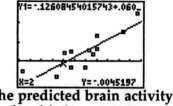

Bring up item 1 in the CALC screen (2nd TRACE).	Type the distress score of X = 2 and press ENTER.	The predicted brain activity level is Y = –0.00452.

Alternately, we can access the **Y1** function from the **Home** screen. To do so, press **VARS**, arrow right to **Y–VARS**, enter **1** for **Function**, enter **1** for **Y1**, then enter the command **Y1(2)**.

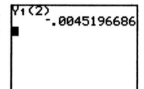

We can also verify that the point $(\bar{x}, \bar{y})$ is on the regression line; but first we must compute the statistics. We can do so simultaneously with the **2–Var Stats** command from the **STAT CALC** menu because the two data sets have the same size. Enter the command **2–Var Stats L1,L2**. Then enter **Y1($\bar{x}$)** by recalling $\bar{x}$ from the **VARS Statistics** menu.

Compute 2–Var Stats.	View statistics.	Y1($\bar{x}$) = $\bar{y}$

(c) With the calculator's diagnostics turned on, the **LinReg(a+bx)** command also displays the values of r and r^2. In this case, $r^2 \approx 0.7713$. Thus, 77.13% of the variation in brain activity among these subjects is explained by the straight-line relationship with social distress score.

Exercise 2.53 Compute the mean and standard deviation of the metabolic rates and mean body masses in Exercise 2.13 and the correlation between these two variables. Use these values to find the equation of the regression line of metabolic rate on lean body mass.

Solution. We first enter the data into lists. Here we use list **L3** for the body masses x and list **L4** for the metabolic rates y. We then enter the command **2–Var Stats L3,L4** to compute the basic statistics.

Enter data.	Compute 2–Var Stats.	View statistics.

We see that $\bar{x} \approx 46.7421$, $S_x \approx 8.28441$, $\bar{y} \approx 1369.5263$, and $S_y \approx 257.5041$. Next, we compute the correlation with the command **LinReg(a+bx) L3,L4**, which also gives us the equation of the regression line $y = a + bx$ of metabolic rate on lean body mass. Finally, we can verify that $b = r \times S_y / S_x \approx 26.87857$ and $a = \bar{y} - b\bar{x} \approx 113.1654$.

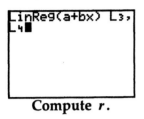

Compute r.

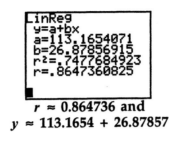

$r \approx 0.864736$ and
$y \approx 113.1654 + 26.87857$

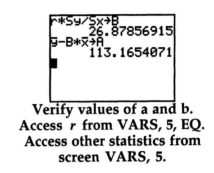

Verify values of a and b.
Access r from VARS, 5, EQ.
Access other statistics from
screen VARS, 5.

2.4 Cautions about Correlation and Regression

We now complete an exercise to demonstrate how to work with the residuals of a least-squares regression line.

Exercise 2.74 The following table gives the speeds (in feet per second) and the mean stride rates for some of the best female American runners.

Speed	15.86	16.88	17.50	18.62	19.97	21.06	22.11
Stride Rate	3.05	3.12	3.17	3.25	3.36	3.46	3.55

(a) Make a scatterplot with speed on the x axis and stride rate on the y axis.
(b) Compute and graph the equation of the regression line of stride rate on speed.
(c) For each of the speeds given, calculate the predicted stride rate and the residual. Verify that the residuals sum to 0.
(d) Plot the residuals against speed.

Solution. (a) We enter speeds into list **L1**, the corresponding stride rates into list **L2**, adjust the **WINDOW** and **STAT PLOT** settings, and graph.

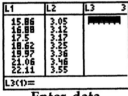

Enter data.

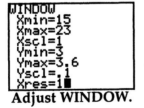

Adjust WINDOW.

Adjust STAT PLOT.

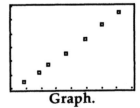

Graph.

(b) Compute the regression line by entering the command **LinReg(a+bx) L1,L2**, then enter the equation of the line into the **Y=** screen and re-graph.

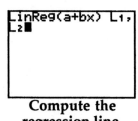 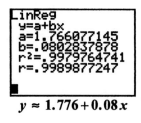 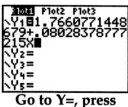 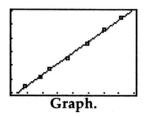

| Compute the regression line. | $y \approx 1.776 + 0.08x$ | Go to Y=, press VARS, then 5, arrow right to EQ, press 1 to enter the line. | Graph. |

(c) Once the regression line is entered into **Y1**, we can access this function from the **VARS, Y–VARS, Function** menu in order to evaluate the predicted stride rates. On the **Home** screen, enter the command **Y1(L1)➔L3** to store the predicted values into list **L3**. Then enter the command **L2 – L3 ➔L4** to compute the residuals and store them into list **L4**.

   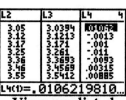

| Press VARS, arrow right to Y–VARS, press 1 for Function, then press 1 for Y1. | Compute predicted values and store into list L3. | Compute residuals and store into L4. | View predicted values and residuals. |

To verify that the residuals sum to 0 (up to round-off error), simply compute the statistics on their values in list **L4** by entering the command **1–Var Stats L4**. We see that their sum, Σx, is essentially 0.

 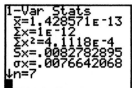

(d) Finally, we make a scatterplot of **L1** versus **L4** to plot the residuals against speed.

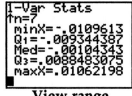   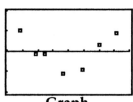

| View range of residuals. | Adjust Y range in the WINDOW. | Adjust Ylist in the STAT PLOT. | Graph. |

CHAPTER
3

```
{1,1,0,1,0,0,1,0
,1,0,0,1,0,1,0,1
,0,0,1,1,1,0,1}
{10,36,9,7,31,24
,11,18,26,29,17,
14,22,4,10,32,5}
```

Producing Data

3.1 First Steps
3.2 Design of Experiments
3.3 Sampling Design
3.4 Toward Statistical Inference

Introduction

In this chapter, we use the TI-83 Plus to simulate the collection of random samples. We also provide a supplementary program that can be used to draw a random sample from a list of integers numbered from *m* to *n*.

3.1 First Steps

In this section, we demonstrate how to generate count data, or *Bernoulli trials,* for a specified proportion p. The data simulates observational "Yes/No" outcomes obtained from a random survey. To generate the data, we use the **randBin** command from the **MATH PRB** menu.

Example Suppose that 62% of students hold a part-time or full-time job at a particular university. Simulate a random survey of 200 students and determine the sample proportion of those who have a job.

Solution. To generate a random list of 1 and 0 responses ("Yes/No"), enter the command **randBin(1, p, n)➧L1**, where p is the specified proportion and n is the desired sample size. Here, use **randBin(1, .62, 200)➧L1**. Then enter the command **1–Var Stats L1**.

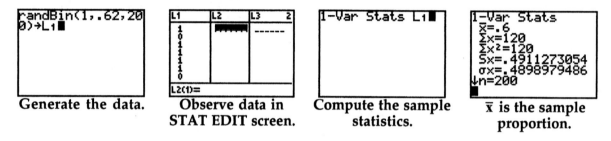

| Generate the data. | Observe data in STAT EDIT screen. | Compute the sample statistics. | $\bar{x}$ is the sample proportion. |

Example Generate 150 observations from a $N(100, 15)$ distribution. Compute the sample statistics to compare $\bar{x}$ with 100 and to compare s with 15.

Solution. Enter the command **randNorm(100, 15, 150)➧L1** and then compute the sample statistics.

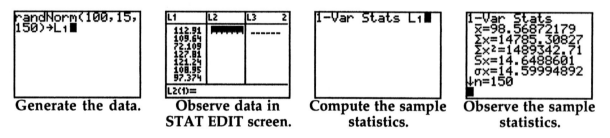

| Generate the data. | Observe data in STAT EDIT screen. | Compute the sample statistics. | Observe the sample statistics. |

3.2 Design of Experiments

Next, we provide a supplementary program that can be used to choose subjects at random from an enumerated group.

The RANDOM Program

```
PROGRAM:RANDOM                    :L₁(K)→L₂(K)
:Disp "LOWER BOUND"               :1+K→K
:Input M                          :End
:Disp "UPPER BOUND"               :A→K
:Input N                          :While K≤N-M+1-I
:Disp "CHOOSE HOW MANY?"          :L₁(K+1)→L₂(K)
:Input R                          :1+K→K
:ClrList L₃                       :End
:seq(J,J,M,N)→L₁                  :L₂→L₁
:For(I,1,R)                       :End
:ClrList L₂                       :L₃→L₁
:randInt(1,N-M+2-I)→A             :ClrList L₂,L₃
:L₁(A)→L₃(I)                      :ClrHome
:1→K                              :Output(1,2,L₁)
:While K<A
```

The **RANDOM** program can be used to choose a random subset of k subjects from a group that has been numbered from m to n. It also can be used to permute an entire set of n subjects so that the group can be assigned randomly to blocks. The program displays the random choices and also stores the values into list **L1**.

Exercise 3.13 Randomly assign 36 subjects into four groups of size 9.

Solution. We execute the **RANDOM** program by numbering the subjects from 1 to 36 and choosing all 36.

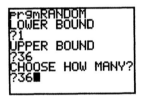

The 36 subjects have been permuted so that they can be assigned randomly to four groups. Simply use consecutive groups of nine for these groups: {16, 8, 18, 35, 17, 2, 24, 28, 5}, {4, 11, 25, 30, 1, 13, 31, 26, 36}, {27, 23, 34, 15, 20, 22, 7, 10, 14}, {32, 9, 21, 3, 33, 19, 6, 12, 29}.

Exercise 3.20 Randomly choose 20 subjects from a group of 40.

Solution. We label the subjects from 1 to 40 and then randomly choose 20.

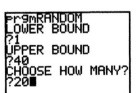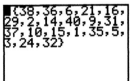

3.3 Sampling Design

The **RANDOM** program can be used to choose a simple random sample from a designated population. This program chooses the sample all at once without repeated choices. But instead, we may want to use a systematic random sample by drawing one subject at a time from sequential groups. The following exercise demonstrates this process.

Exercise 3.47 Choose a systematic random sample of four addresses from a list of 100.

Solution. Because the list of 100 divides evenly into four groups of 25, we will choose one address from each of the groups 1–25, 26–50, 51–75, and 76–100. And because we are choosing only one number at a time, we can use the **randInt(** command from the **MATH PRB** menu. To choose one integer from a to b, enter the command **randInt(a,b)**. We use the command four times as shown below to obtain the desired sample.

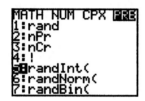 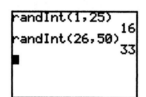 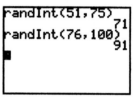

Number of Ways to Choose

When choosing a simple random sample of size r from a population of size n, we generally choose without repeats and without regard to order. Such a choice is called a *combination*. The number of possible combinations (often called "n choose r") can be computed with the **nCr** button from the **MATH PRB** menu.

Exercises 3.44, 3.46 (a) How many ways are there to choose five blocks from a group of blocks labeled 1–44? (b) How many ways are there to choose a stratified sample of five blocks so that there is one chosen from blocks 1–6, two chosen from blocks 7–18, and three chosen from blocks 19–44. (c) In each case, choose such a sample.

Solution. (a) There are "44 choose 5" ways to pick five blocks at random from 44, which is computed by **44 nCr 5** on the calculator. Thus, there are 1,086,008 possible samples in this case.

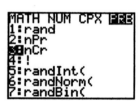

(b) We choose one from the first group of 6 blocks, choose two from the second group of 12 blocks, and choose three from the third group of 26 blocks. The total number of ways to choose in this manner is given by (6 nCr 1)×(12 nCr 2)×(26 nCr 3) = 6 × 66 × 2600 = 1,029,600.

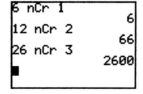 

(c) We can choose the samples in each case with the **RANDOM** program.

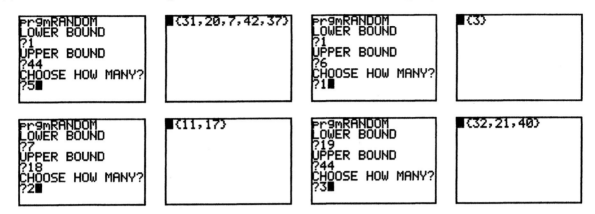

3.4 Toward Statistical Inference

In Section 3.1, we used the command **randBin(1, *p*, *n*)▸L1** to generate a random sample of "Yes/No" responses. In this section, we will demonstrate how to simulate the collection of multiple samples. In particular, we are concerned with the total number of "Yes" responses, the sample proportion for each sample, and the resulting average of all sample proportions. This simulation also can be made with the **randBin(** command from the **MATH PRB** menu.

Exercise 3.73 (a) We have a coin for which the probability of heads is 0.60. We toss the coin 25 times and count the number of heads in this sample. Then we repeat the process for a total of 50 samples of size 25. Simulate the counts of heads for these 50 samples of size 25, compute the sample proportion for each sample, and make a histogram of the sample proportions.

Solution. One simulated sample of counts can be obtained with the command **randBin(25, 0.6)**. But because we want 50 samples of size 25, we will use the command **randBin(25, 0.6, 50)▸L1** in order to generate the counts and store them into list **L1**. Then the command **L1/25 ▸L2** will compute the sample proportion for each sample and store the results in list **L2**. By computing the sample statistics on list **L2**, we obtain the average of all the sample proportions.

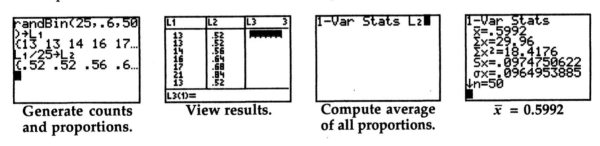

| Generate counts and proportions. | View results. | Compute average of all proportions. | $\bar{x}$ = 0.5992 |

By observing the list of sample proportions in **L2**, we see that we may never have a sample proportion that equals the real proportion of 0.60. However, the *average* of all 50 sample proportions was 0.5992, which is very close to 0.60. Lastly, adjust the **WINDOW** and **STAT PLOT** settings to see a histogram of the sample proportions in list **L2**.

CHAPTER
4

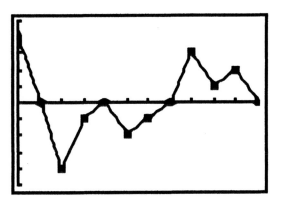

Probability: The Study of Randomness

4.1 Randomness
4.2 Probability Models
4.3 Random Variables
4.4 Means and Variances
    of Random Variables
4.5 General Probability

Introduction

In this chapter, we show how to use the TI-83 Plus to generate some random sequences. We then see how to make a probability histogram for a discrete random variable and how to compute its mean and standard deviation. We conclude with a program for the Law of Total Probability and Bayes' Rule.

4.1 Randomness

In this section, we work some exercises that use the TI-83 Plus to generate various random sequences. We shall need the **randBin(** and **randInt(** commands from the **MATH PRB** menu.

Exercise 4.5 Simulate 100 free throws shot independently by a player who has 0.5 probability of making a single shot. Examine the sequence of hits and misses.

Solution. The command **randBin(1,.5,100)▶L1** will generate and store a list of 100 "1's and 0's" to represent the hits and misses. The results are stored in the **STAT EDIT** screen.

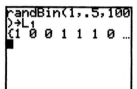
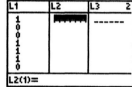

Exercise 4.7 Simulate rolling four fair dice over and over again. What percentage of the time was there at least one "6" in the set of four rolls?

Solution. The command **randInt(j, k)** generates a random integer from *j* to *k*. The command **randInt(j, k, n)** generates *n* such random integers. Here we enter the command **randInt(1, 6, 4)** to simulate four rolls of dice numbered 1 to 6. After entering the command once, keep pressing **ENTER** to reexecute the command.

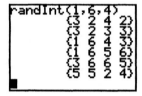
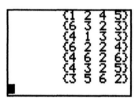
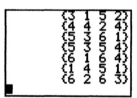
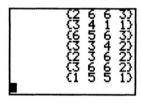

In the 27 sets shown above, there are 14 sets with at least one "6," which gives 51.85%.

Exercise 4.9 Simulate 100 binomial observations each with $n = 20$ and $p = 0.3$. Convert the counts into percents and make a histogram of these percents.

Solution. The command **randBin(20, 0.3,100)▶L1** will generate the observations and put them in list **L1**, and the command **100*L1/20▶L2** will put the percents into list **L2**.

Generate counts
and percents.

View results.

Next, adjust the **WINDOW** and **STAT PLOT** settings to make a histogram of the percents in list **L2**.

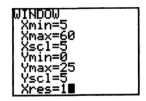 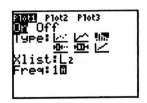 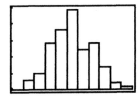

4.2 Probability Models

In this section, we demonstrate some of the basic concepts of probability models.

Example 4.6 (a) Generate random numbers between 0 and 1. (b) Make a histogram of 100 such randomly generated numbers.

Solution. (a) Simply enter the command **rand** from the **MATH PRB** menu. After the command has been entered once, keep pressing **ENTER** to continue generating more random values between 0 and 1. (b) Enter the command **rand(100)➔L1** to generate and store 100 random numbers between 0 and 1.

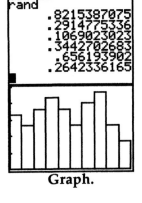

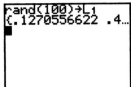
Generate list.

Adjust WINDOW.   Adjust STAT PLOT.   Graph.

Example 4.9 The first digit V of numbers in legitimate records often follow the distribution given in the table below, known as Benford's Law. (a) Verify that the table defines a legitimate probability distribution. (b) Compute the probability that the first digit is 6 or greater.

First digit V	1	2	3	4	5	6	7	8	9
Probability	0.301	0.176	0.125	0.097	0.079	0.067	0.058	0.051	0.046

Solution. (a) To sum these probabilities, we can enter the values into a list and use the two commands **sum(** and **seq(** from the **LIST MATH** and **LIST OPS** menus.

First, enter the values of the digits into list **L1** (optional) and enter the probabilities into list **L2**. To verify that the table gives a legitimate probability distribution, enter the command **sum(seq(L2(I), I, 1, 9))**, which sums the values in list **L2** as the index **I** ranges from 1 to 9.

Enter distribution.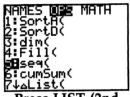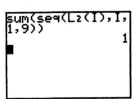

Enter distribution.

Press LIST (2nd STAT), arrow right to MATH, press 5.

Press LIST (2nd STAT), arrow right to OPS, press 5.

Sum the list from 1 to 9.

(b) After summing all the probabilities, press **2nd ENTER** to recall the previous command, then edit it to **sum(seq(L2(I), I, 6, 9))** to sum the sixth through ninth probabilities. We see that the probability of the first digit being 6 or greater is 0.222.

4.3   Random Variables

In this section, we work exercises that compute various probabilities involving random variables. We begin though with an exercise on constructing a probability histogram.

Exercises 4.43, 4.45 The table below gives the distributions of rooms for owner-occupied units and for renter-occupied units in San Jose, California. Make probability histograms of these two distributions. If X represents the number of rooms in a randomly chosen owner-occupied unit, compute $P(X > 5)$.

Rooms	1	2	3	4	5	6	7	8	9	10
Owned	0.003	0.002	0.023	0.104	0.210	0.224	0.197	0.149	0.053	0.035
Rented	0.008	0.027	0.287	0.363	0.164	0.093	0.039	0.013	0.003	0.003

Solution. We enter the values of the rooms into list **L1**, the owner probabilities into list **L2**, and the renter probabilities into list **L3**. Then we make separate histograms for an **Xlist** of **L1** with frequencies of either **L2** or **L3**.

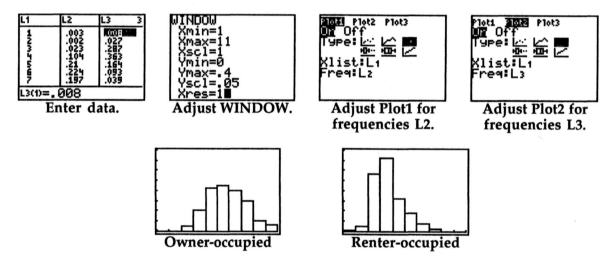

Enter data.	Adjust WINDOW.	Adjust Plot1 for frequencies L2.	Adjust Plot2 for frequencies L3.

Owner-occupied	Renter-occupied

The value $P(X > 5)$ is equivalent to $P(6 \le X \le 10)$. This value can be computed with the command **sum(seq(L2(I), I, 6, 10))**, which gives 0.658.

Exercise 4.55 Let $Y \sim U[0, 2]$. (a) Graph the density curve.   (b) Find $P(0.5 \le Y \le 1.3)$.

Solution. For $Y \sim U[0, 2]$, the height of the density curve is $1/(2 - 0) = 0.5$. We simply enter this function into the **Y=** screen and use item 7 from the **CALC** menu to compute the area between the values 0.5 and 1.3.

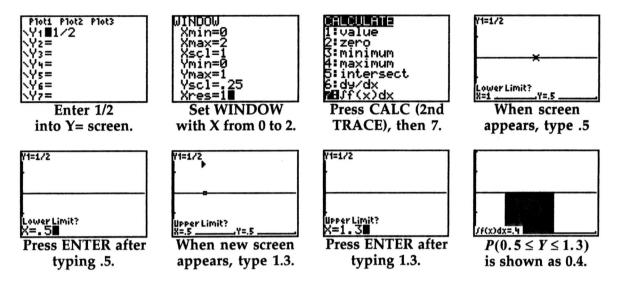

| Enter 1/2 into Y= screen. | Set WINDOW with X from 0 to 2. | Press CALC (2nd TRACE), then 7. | When screen appears, type .5 |

| Press ENTER after typing .5. | When new screen appears, type 1.3. | Press ENTER after typing 1.3. | $P(0.5 \leq Y \leq 1.3)$ is shown as 0.4. |

We conclude this section by working an exercise using the normal density curve that reviews the normal distribution calculations from Section 1.3.

Exercise 4.58   After an election in Oregon, voter records showed that 56% of registered voters actually voted. A survey of 663 registered voters is conducted and the sample proportion $\hat{p}$ of those who claim to have voted is obtained. For all random samples of size 663, these values of the sample proportions $\hat{p}$ will follow an approximate normal distribution with mean $\mu = 0.56$ and standard deviation $\sigma = 0.019$. Use this distribution to compute $P(0.52 \leq \hat{p} \leq 0.60)$ and $P(\hat{p} \geq 0.72)$.

Solution. We use the built-in **normalcdf(** command from the **DISTR** menu. For $\hat{p} \sim N(0.56, 0.019)$, we use $P(0.52 \leq \hat{p} \leq 0.60)$ = **normalcdf(.52, 0.6, .56, .019)** and $P(\hat{p} \geq 0.72)$ = **normalcdf(.72, 1E99, .56, .019)**.

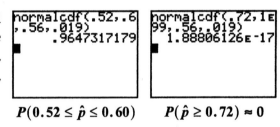

$P(0.52 \leq \hat{p} \leq 0.60)$ $P(\hat{p} \geq 0.72) \approx 0$

4.4   Means and Variances of Random Variables

We now show how to compute the mean and standard deviation of a discrete random variable for which the range of measurements and corresponding probabilities are given.

Exercise 4.61   The table below gives the distributions of the number of rooms for owner-occupied units and renter-occupied units in San Jose, California. Calculate the mean and the standard deviation of the number of rooms for each type.

Rooms	1	2	3	4	5	6	7	8	9	10
Owned	0.003	0.002	0.023	0.104	0.210	0.224	0.197	0.149	0.053	0.035
Rented	0.008	0.027	0.287	0.363	0.164	0.093	0.039	0.013	0.003	0.003

Solution. Enter the measurements (rooms) into list **L1** and the probabilities into lists **L2** and **L3**. For the owner-occupied units, enter the command **1–Var Stats L1,L2**. We see that the average number of rooms for owner-occupied units is $\mu = 6.284$ with a standard deviation of $\sigma \approx 1.64$. For the renter-occupied units, enter the command **1–Var Stats L1,L3** to obtain $\mu = 4.187$ and $\sigma \approx 1.3077$.

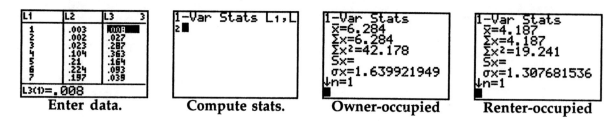

| Enter data. | Compute stats. | Owner-occupied | Renter-occupied |

Sampling from a Discrete Distribution

Previously, we have used commands such as **randNorm(100, 15, *n*)**►**L1**, **2*rand(*n*)►L1**, and **randBin(20, 0.3, 100)►L1** to generate lists of random values from normal, uniform, and binomial distributions. We now provide a short program that will draw a random sample from a discrete distribution provided its table of probabilities is given.

The DISTSAMP Program

```
PROGRAM:DISTSAMP          :End
:Disp "NO. OF POINTS?"    :L₁(J)→L₄(I)
:Input N                  :End
:ClrList L₄               :1-Var Stats L₁,L₂
:cumSum(L₂)→L₃            :x̄→B
:For(I,1,N)               :1-Var Stats L₄
:1→J                      :x̄→C
:rand→A                   :ClrHome
:While A>L₃(J)            :Disp "REAL MEAN",B
:J+1→J                    :Disp "SAMPLE MEAN",C
```

Example Draw a random sample of 250 points according to a distribution that follows Benford's Law as given in the following table. Compare the sample mean with the true average, make a histogram of the sample points, and find the sample proportions of each digit.

First digit V	1	2	3	4	5	6	7	8	9
Probability	0.301	0.176	0.125	0.097	0.079	0.067	0.058	0.051	0.046

Solution. Before running the **DISTSAMP** program, enter the values of the digits into list **L1** and enter the probabilities into list **L2**. Then bring up the program and enter the desired number of sample points. This sample will be stored in list **L4**.

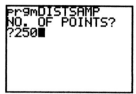

Enter distribution.   Execute program.

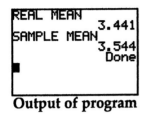
Output of program

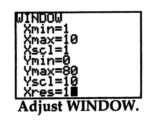
Adjust WINDOW.

Adjust STAT PLOT
for a histogram of L4.

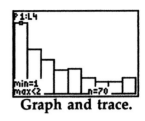
Graph and trace.

By graphing and tracing the histogram of sample points in list **L4**, we can see the sample counts of each digit. In this case, the digit 1 was chosen 70 times for a sample proportion of $70/250 = 0.28$.

Mean and Standard Deviation of an Independent Sum

Here we show how to verify the mean and standard deviation of a random variable of the form $Z = aX + bY$, where X and Y are independent.

Example Suppose that $Z = 0.2X + 0.8Y$, where $\mu_X = 5$, $\sigma_X = 2.9$, $\mu_Y = 13.2$, and $\sigma_Y = 17.6$. Assuming that X and Y are independent, find the mean and standard deviation of Z.

Solution. We can think of X as taking two values $5 - 2.9 = 2.1$ and $5 + 2.9 = 7.9$ (see Exercise 4.78 in the text). Likewise, we can consider Y to assume only the values $13.2 - 17.6 = -4.4$ and $13.2 + 17.6 = 30.8$. We enter these values of X and Y into lists **L1** and **L2**; however, we list each X value twice in consecutive fashion (2.1, 2.1, 7.9, 7.9), and we list the Y values twice in alternating fashion (−4.4, 30.8, −4.4, 30.8). Next, we enter the command **0.2*L1 + 0.8*L2➜L3** to send the possible values of Z to list **L3**. Finally, enter **1–Var Stats L3** to compute the mean and standard deviation of Z.

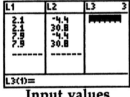
Input values
of X and Y.

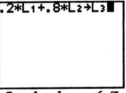
Send values of Z
to list L3.

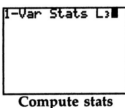
Compute stats
on L3.

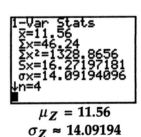
$\mu_Z = 11.56$
$\sigma_Z \approx 14.09194$

We thereby can verify that $\mu_Z = 0.2\mu_X + 0.8\mu_Y$ and, because X and Y are independent, that $\sigma_Z = \sqrt{0.2^2 \sigma_X^2 + 0.8^2 \sigma_Y^2}$. However, if $Z = aX + bY$ and X and Y are *not* independent, then $\sigma_Z^2 = a^2\sigma_X^2 + b^2\sigma_Y^2 + 2\rho a\sigma_X b\sigma_Y$ (see Example 4.28 in the text).

4.5 General Probability

We conclude this chapter with a short program for the Law of Total Probability and Bayes' Rule. The **BAYES** program given below computes and displays the total probability $P(C)$ according to the formula

$$P(C) = P(A_1)P(C \mid A_1) + P(A_2)P(C \mid A_2) + \ldots + P(A_n)P(C \mid A_n)$$

Before executing the **BAYES** program, enter the given probabilities $P(A_1), \ldots, P(A_n)$ into list **L1** and the given conditionals $P(C \mid A_1), \ldots, P(C \mid A_n)$ into list **L2**.

The program also stores the probabilities of the intersections $P(C \cap A_1), \ldots, P(C \cap A_n)$ in list **L3**, and stores the reverse conditionals $P(A_1 \mid C), \ldots, P(A_n \mid C)$ in list **L4**. Finally, the conditional probabilities $P(A_1 \mid C'), \ldots, P(A_n \mid C')$ are stored in list **L5**, and the conditional probabilities $P(C \mid A_1'), \ldots, P(C \mid A_n')$ are stored in list **L6**. To help the user keep track of which list contains which probabilities, the program displays a description.

The BAYES Program

```
PROGRAM:BAYES                              :Disp "TOTAL PROB"
:sum(seq(L₁(I)*L₂(I),I,1,dim(L₁)))→T       :Disp round(T,4)
:L₁*L₂→L₃                                   :Disp "C AND As : L₃"
:L₃/T→L₄                                    :Disp "As GIVEN C : L₄"
:L₁*(1-L₂)/(1-T)→L₅                         :Disp "As GIVEN C' : L₅"
:T*(1-L₄)/(1-L₁)→L₆                         :Disp "C GIVEN A's : L₆"
```

Exercise 4.104 The voters in a large city are 40% white, 40% black, and 20% Hispanic. A mayoral candidate expects to receive 30% of the white vote, 90% of the black vote, and 50% of the Hispanic vote. Apply the **BAYES** program to compute the percent of the overall vote that the candidate expects, and to analyze the other computed conditional probabilities.

Solution. Here we let A_1 = white voters, A_2 = black voters, and A_3 = Hispanic voters. We enter the probabilities of these events into list **L1**. We let C be the event that a person votes for the candidate. Then $P(C \mid A_1) = 0.30$, $P(C \mid A_2) = 0.90$, and $P(C \mid A_3) = 0.50$, and we enter these conditional probabilities into list **L2**. Then we execute the **BAYES** program.

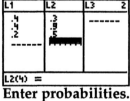

Enter probabilities.
Execute BAYES.

Output of program

Probability of
intersections in L3

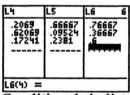

Conditionals in lists
L4, L5, and L6

We first see that the candidate can expect to receive 58% of the overall vote. This result is obtained by $P(C) = 0.4 \times 0.3 + 0.4 \times 0.9 + 0.2 \times 0.5 = 0.58$. The other computed probabilities are also stored in the designated lists.

List **L3** contains the probabilities of the intersections. The probability that a voter is white and will vote for the candidate is $P(A_1 \cap C) = 0.12$. The probability that a voter is black and will vote for the candidate is $P(A_2 \cap C) = 0.36$. The probability that a voter is Hispanic and will vote for the candidate is $P(A_3 \cap C) = 0.1$. These values are obtained with the formula $P(A_i \cap C) = P(A_i) \times P(C \mid A_i)$, and are found by multiplying the terms in lists **L1** and **L2**.

List **L4** contains the reverse conditional probabilities of being white, black, Hispanic given that one *will* vote for the candidate. This list is the direct application of Bayes' Rule. These conditional probabilities are $P(A_1 \mid C) = 0.2069$, $P(A_2 \mid C) = 0.62069$, and $P(A_3 \mid C) = 0.17241$, respectively. This list is obtained by dividing the respective intersection probabilities in list **L3** by $P(C) = 0.58$.

List **L5** contains the reverse conditional probabilities of being white, black, Hispanic given that one will *not* vote for the candidate. These values are $P(A_1 \mid C') = 0.66667$, $P(A_2 \mid C') = 0.09524$, and $P(A_3 \mid C') = 0.2381$, respectively.

List **L6** contains the conditional probabilities of voting for the candidate given that a voter is *not* white, *not* black, and *not* Hispanic. These values are $P(C \mid A'_1) = 0.76667$, $P(C \mid A'_2) = 0.36667$, and $P(C \mid A'_3) = 0.6$, respectively.

We note that given any conditional probability $P(C \mid D)$, then the complement conditional probability is given by $P(C' \mid D) = 1 - P(C \mid D)$. Thus, lists for complement conditional probabilities do not need to be generated. For example, the respective conditional probabilities of *not* voting for the candidate given that one is white, black, and Hispanic are respectively 0.70, 0.10, and 0.50. These values are the complement probabilities of the originally given conditionals.

Exercise 4.111 Cystic fibrosis. The probability that a randomly chosen person of European ancestry carries an abnormal CF gene is 1/25. If one is a carrier of this gene, then a test for it will be positive 90% of the time. If a person is not a carrier, then the test will never be positive. Jason tests positive. What is the probability that he is a carrier?

Solution. We let C be the event that a person of European ancestry tests positive, let A_1 be carriers of the gene, and let A_2 be non-carriers. Because $P(A_1) = 1/25$, then $P(A_2) = 24/25$. Also, $P(C \mid A_1) = 0.90$, $P(C \mid A_2) = 0$. We are trying to compute $P(A_1 \mid C)$, which is the probability of being a carrier given that one has tested positive.

We enter the probabilities $P(A_1)$ and $P(A_2)$ into list **L1**, enter the conditional probabilities $P(C \mid A_1)$ and $P(C \mid A_2)$ into list **L2**, and execute the **BAYES** program.

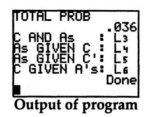 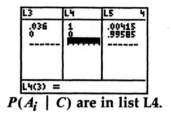

Enter probabilities.   **Output of program** $P(A_i \mid C)$ are in list **L4.**
Execute BAYES.

List **L4** contains the conditionals $P(A_i \mid C)$. From the first entry, we see that $P(A_1 \mid C) = 1$. Because Jason has tested positive, there is a 100% chance that he is a carrier. (He must be a carrier because it is impossible for non-carriers to test positive.) This value is also given by

$$P(A_1 \mid C) = \frac{P(A_1 \cap C)}{P(C)} = \frac{(1/25) \times 0.90}{(1/25) \times 0.90 + (24/25) \times 0} = 1$$

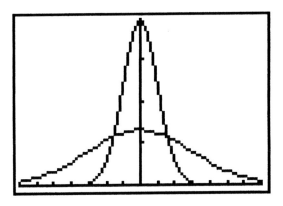

Sampling Distributions

Introduction

In this chapter, we show how to compute probabilities involving a binomial distribution and probabilities involving the sample mean $\bar{x}$.

5.1 Sampling Distributions for Counts and Proportions

We begin by demonstrating how to compute various probabilities for a given binomial distribution. To do so, we will need the **binompdf(** and **binomcdf(** commands (items 0 and A) from the **DISTR** menu.

Binomial Probabilities

For a binomial distribution $X \sim B(n,p)$, we compute the probability of exactly k successes, $P(X = k)$, by entering the command **binompdf(n, p, k)**. The probability $P(X \le k) = P(0 \le X \le k)$ of at most k successes is computed with the command **binomcdf(n, p, k)**. The probability of there being at least k successes is given by $P(X \ge k) = 1 - P(X \le k-1)$, and is computed with the command **1 − binomcdf(n, p, k−1)**. The following three examples demonstrate these calculations.

Example 5.4 Let $X \sim B(150, 0.08)$. Calculate $P(X = 10)$ and $P(X \le 10)$.

Solution. Simply enter the commands **binompdf(150, .08, 10)** and **binomcdf(150, .08, 10)** to obtain $P(X = 10) \approx 0.106959$ and $P(X \le 10) \approx 0.338427$.

Example 5.5 Let $X \sim B(15, 0.08)$. Make a probability table and probability histogram of the distribution. Also make a table of the cumulative distribution and use it to find $P(X \le 1)$.

Solution. Because there are $n = 15$ attempts, the possible number of successes range from 0 to 15. So we first enter the integer values $0, 1, \ldots, 15$ into list **L1**. We can do so directly or we can use the command **seq(K, K, 0, 15)➔L1**. Next, we use the commands **binompdf(15, .08)➔L2** and **binomcdf(15, .08)➔L3** to enter the probability distribution values $P(X = k)$ into list **L2** and the cumulative distribution values $P(X \le k)$ into list **L3**. From the values in list **L3**, we can see that $P(X \le 1) \approx 0.65973$.

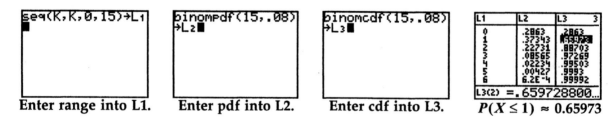

| Enter range into L1. | Enter pdf into L2. | Enter cdf into L3. | $P(X \le 1) \approx 0.65973$ |

To see a probability histogram of the distribution, adjust the **WINDOW** and **STAT PLOT** settings for a histogram of **L1** with frequencies **L2**. We note that the probabilities beyond $X = 6$ will not be observable.

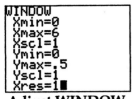
Adjust WINDOW.

Adjust STAT PLOT.

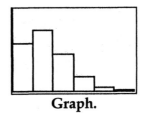
Graph.

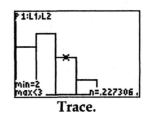
Trace.

Example 5.6 Let $X \sim B(12, 0.25)$, compute $P(X \geq 5)$.

Solution. We use the probability of the complement to obtain $P(X \geq 5) = 1 - P(X \leq 4)$, which is computed by **1 – binomcdf(12, .25, 4)** ≈ 0.15764.

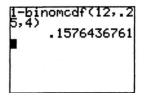

Probabilities for $\hat{p}$

The next two problems show how to make probability calculations for a sample proportion $\hat{p}$ by converting to a binomial probability.

Example 5.8 Suppose that 60% of all adults agree that they like shopping for clothes, but often find it frustrating and time-consuming. In a nationwide sample of 2500 adults, let $\hat{p}$ be the sample proportion of adults who agree with this response. Compute $P(\hat{p} \geq 0.58)$.

Solution. Because 58% of 2500 is 1450, we must compute $P(X \geq 1450)$, where $X \sim B(2500, 0.60)$. Instead, we may compute $1 - P(X \leq 1499)$ by **1 – binomcdf(2500, .60, 1499)** ≈ 0.98018.

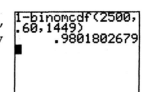

Exercise 5.15 (b) For an SRS of size $n = 1011$ and assuming a true proportion of $p = 0.06$, what is the probability that a sample proportion $\hat{p}$ lies between 0.05 and 0.07?

Solution. For $n = 1011$ and $p = 0.06$, then $P(0.05 \leq \hat{p} \leq 0.07) = P(n \times 0.05 \leq X \leq n \times 0.07) = P(50.55 \leq X \leq 70.77) = P(51 \leq X \leq 70) = P(X \leq 70) - P(X \leq 50)$, where $X \sim B(1011, 0.06)$. We find this value by entering **binomcdf(1011, .06, 70) – binomcdf(1011, .06, 50)**, and obtain a probability of about 0.815266.

Normal Approximations

We conclude this section by showing how to approximate a sample proportion probability and a binomial probability with a normal distribution.

Example 5.10 With $n = 2500$ and $p = 0.60$ as in Example 5.8, use the approximate distribution of $\hat{p}$ to estimate $P(\hat{p} \geq 0.58)$.

Solution. The distribution of $\hat{p}$ is approximately normal with $\mu = p = 0.60$ and $\sigma = \sqrt{p(1-p)/n} = \sqrt{0.60 \times 0.40 / 2500} \approx 0.0098$. Thus, $P(\hat{p} \geq 0.58) \approx P(Y \geq 0.58)$, where $Y \sim N(0.60, 0.0098)$. The command **normalcdf(.58, 1E99, .6, .0098)** gives a probability of about 0.97936.

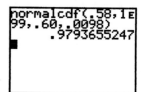

Exercise 5.20 Let $X \sim B(1500, 0.7)$. (a) What are the mean and standard deviation of X? (b) Use the normal approximation to find $P(X \geq 1000)$.

Solution. (a) The mean is $\mu = np = 1500 \times 0.70 = 1050$, and the standard deviation is $\sigma = \sqrt{np(1-p)} = \sqrt{1500 \times 0.70 \times 0.30} = \sqrt{315}$.

(b) We now let $Y \sim N(1050, \sqrt{315})$. Then $P(X \geq 1000) \approx P(Y \geq 1000)$, which is found with the command **normalcdf(1000,1E99,1050,√(315))**. We see that $P(X \geq 1000) \approx 0.9976$.

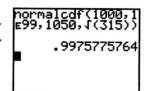

5.2 The Sampling Distribution of a Sample Mean

We now show how to compute various probabilities involving the sample mean $\bar{x}$. To do so, we make use of the fact that for random samples of size n from a $N(\mu, \sigma)$ distribution, the sample mean $\bar{x}$ follows a $N(\mu, \sigma/\sqrt{n})$ distribution.

Exercise 5.37 Sheila's glucose level one hour after ingesting a sugary drink varies according to the normal distribution with $\mu = 125$ mg/dl and $\sigma = 10$ mg/dl.

(a) If a single glucose measurement is made, what is the probability that Sheila measures above 140?
(b) What is the probability that the sample mean from four separate measurements is above 140?

Solution. (a) We compute $P(X > 140)$ for $X \sim N(125, 10)$ with the command **normalcdf(140, 1E99, 125, 10)**. Then, $P(X > 140) \approx 0.0668$.

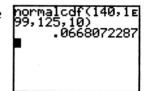

(b) For an SRS of size $n = 4$, $\bar{x}$ has a mean of $\mu = 125$ and a standard deviation of $\sigma / \sqrt{n} = 10/\sqrt{4} = 5$. So now we compute $P(\bar{x} > 140)$ for $\bar{x} \sim N(125, 5)$ and obtain a value of about 0.00135.

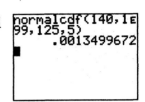

Exercise 5.39 Sheila's glucose level one hour after ingesting a sugary drink varies according to the normal distribution with $\mu = 125$ mg/dl and $\sigma = 10$ mg/dl. What is the level L such that there is only 0.05 probability that the mean glucose level of four test results falls above L for Sheila's glucose level distribution?

Solution. As in Exercise 5.37, $\bar{x} \sim N(125, 5)$. So we must find the inverse normal value L for which $P(\bar{x} > L) = 0.05$ or, equivalently, $P(\bar{x} \le L) = 0.95$. We compute this value with the **invNorm(** command from the **DISTR** menu by entering **invNorm(.95,125,5)**. We see that only about 5% of the time should $\bar{x}$ be larger than $L = 133.224$.

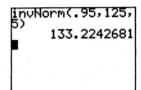

Exercise 5.59 The weight of eggs produced by a certain breed of hen is normally distributed with a mean of 65 g and a standard deviation of 5 g. For random cartons of 12 eggs, what is the probability that the weight of a carton falls between 750 g and 825 g?

Solution. If the total weight of 12 eggs falls between 750 g and 825 g, then the sample mean $\bar{x}$ falls between $750/12 = 62.5$ g and $825/12 = 68.75$ g. So, we compute $P(62.5 \le \bar{x} \le 68.75)$ for $\bar{x} \sim N(65, 5/\sqrt{12})$ by entering the command **normalcdf(62.5, 68.75, 65, 5/√(12))**.

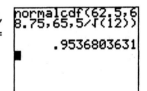

Note: By the Central Limit Theorem, when sampling from an non-normal population with a "large" sample size n, the sample mean $\bar{x}$ follows an *approximate* $N(\mu, \sigma/\sqrt{n})$ distribution. Thus, probabilities involving $\bar{x}$ can be approximated by a normal distribution calculation as in Exercise 5.37 (b) and Exercise 5.59.

Sum of Independent Normal Measurements

Let $X \sim N(\mu_X, \sigma_X)$ and let $Y \sim N(\mu_Y, \sigma_Y)$. Assuming that X and Y are independent measurements, then $X \pm Y$ follows a $N\left(\mu_X \pm \mu_Y, \sqrt{\sigma_X^2 + \sigma_Y^2}\right)$ distribution.

Example 5.19 Suppose $X \sim N(110, 10)$ and $Y \sim N(100, 8)$. If X and Y are independent, then what is the probability that X is less than Y?

Solution. The value $P(X < Y)$ is equivalent to $P(X - Y < 0)$. So we need to use the distribution of the difference $X - Y$ which is $N(110 - 100, \sqrt{10^2 + 8^2}) = N(10, \sqrt{164})$. By entering the command **normalcdf(⁻1E99, 0, 10, √(164))**, we find that $P(X - Y < 0) \approx 0.21744$.

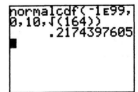

Sum and Difference of Sample Means

Let $\bar{x}$ be the sample mean from an SRS of size n from a $N(\mu_X, \sigma_X)$ distribution, and let $\bar{y}$ be the sample mean from an independent SRS of size m from a $N(\mu_Y, \sigma_Y)$ distribution. Then, the sum/difference $\bar{x} \pm \bar{y}$ follows a $N\left(\mu_X \pm \mu_Y, \sqrt{\sigma_X^2 / n + \sigma_Y^2 / m}\right)$ distribution.

Exercise 5.46 Let $\bar{y}$ be the sample mean from a group of size 30 from a $N(4.8, 1.5)$ population, and let $\bar{x}$ be the sample mean from an independent group of size 30 from a $N(2.4, 1.6)$ population. What is the distribution of $\bar{y} - \bar{x}$? Find $P(\bar{y} - \bar{x} \geq 1)$.

Solution. First, $\bar{y} - \bar{x} \sim N(4.8 - 2.4, \sqrt{1.5^2 / 30 + 1.6^2 / 30}) = N(2.4, 0.4)$. Thus, we find $P(\bar{y} - \bar{x} \geq 1)$ with the command **normalcdf(1, 1E99, 2.4, 0.4)** and obtain a value of 0.999767.

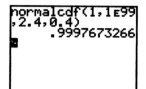

6

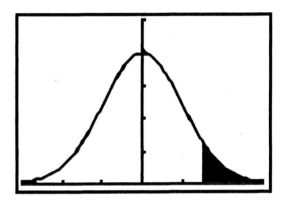

Introduction
to Inference

Introduction

In this chapter, we show how to use the TI-83 Plus to compute confidence intervals and conduct hypothesis tests for the mean μ of a normally distributed population with known standard deviation σ.

6.1 Estimating with Confidence

In this section, we show how to compute a confidence interval for the mean of a normal population with known standard deviation σ. To do so, we will use the built-in **ZInterval** feature (item 7) from the **STAT TESTS** menu. The following two exercises demonstrate how to use this feature with summary statistics and with a data set.

Exercise 6.5 In a study of bone turnover in young women, serum TRAP was measured in 31 subjects and the mean was 13.2 U/l. Assume that the standard deviation is known to be 6.5 U/l. Give the margin of error and find a 95% confidence interval for the mean of all young women represented by this sample.

Solution. Bring up the **ZInterval** screen, set the **Inpt** to **Stats**, then enter the given values of 6.5 for σ, 13.2 for $\bar{x}$, and 31 for n. Enter the desired confidence level, then press **ENTER** on **Calculate**. We obtain a 95% confidence interval of (10.912, 15.488).

Because the confidence interval is of the form $\bar{x} \pm m$, we can find the margin of error m by subtracting $\bar{x}$ from the right endpoint of the interval: $15.488 - 13.2 = 2.288$.

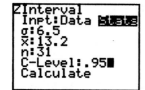
Enter stats into ZInterval screen, then Calculate.

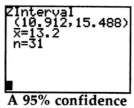
A 95% confidence interval is displayed.

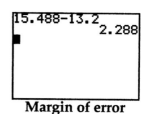
Margin of error

Exercise 6.17 Here are the values of the average speed (in mph) for a sample of trials on a vehicle undergoing a fuel efficiency test. Assume that the standard deviation is 10.3 mph. Estimate the mean speed at which the vehicle was driven with 95% confidence.

21.0	19.0	18.7	39.2	45.8	19.8	48.4	21.0	29.1	35.7
31.6	49.0	16.0	34.6	36.3	19.0	43.3	37.5	16.5	34.5

Solution. First, enter the data into a list, say list **L1**. Next, bring up the **ZInterval** screen, set the **Inpt** to **Data**, and enter the given value of 10.3 for σ. Set the **List** to L1 with frequencies 1, enter the desired confidence level, and press **ENTER** on **Calculate**.

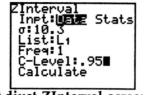
Enter data.

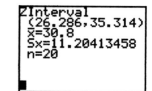
Adjust ZInterval screen.

The confidence interval and statistics are displayed.

Choosing the Sample Size

Suppose we want to find the minimum sample size n that will produce a desired margin of error m with a specific level of confidence. To do so, we can use the formula $n \geq (z* \times \sigma / m)^2$, where $z*$ is the appropriate critical value. We also could use a program that computes the (rounded-up) sample size. To execute the ZSAMPSZE program that follows, we simply input the values of the standard deviation σ, the desired margin of error, and the desired confidence level in decimal.

The ZSAMPSZE Program

```
PROGRAM:ZSAMPSZE          :If int(M)=M
:Disp "STANDARD DEV."     :Then
:Input S                  :M→N
:Disp "DESIRED ERROR"     :Else
:Input E                  :int(M+1)→N
:Disp "CONF. LEVEL"       :End
:Input R                  :ClrHome
:invNorm((R+1)/2,0,1)→Q   :Disp "SAMPLE SIZE="
:(Q*S/E)²→M               :Disp int(N)
```

Example 6.6 Suppose we want a margin of error of $2000 with 95% confidence when estimating the mean debt for students completing their undergraduate studies. The standard deviation is about $49,000. (a) What sample size is required? (b) What sample size would be required to obtain a margin of error of $1500?

Solution. The critical value for 95% confidence is $z* = 1.96$. Using this value in the formula $(z* \times \sigma / m)^2$, with $\sigma = 49,000$ and $m = 2000$, we obtain a necessary sample size of $n = 2306$. The same result is obtained by using the ZSAMPSIZE program. Working part (b) similarly with $m = 1500$, we obtain a required sample size of $n = 4100$.

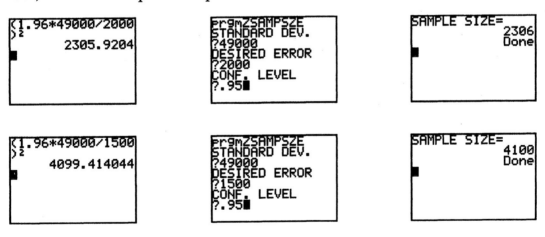

6.2 Tests of Significance

We now show how to use the TI-83 Plus to perform one-sided and two-sided hypothesis tests about the mean μ of a normally distributed population for which the standard deviation σ is known. To do so, we will use the **Z–Test** feature (item 1) from the **STAT TESTS** menu. We can use this feature to work with either summary statistics or data sets.

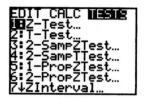

Example 6.14 The mean systolic blood pressure for males 35 to 44 years of age is 128 and the standard deviation is 15. But for a sample of 72 company executives in this age group, the mean systolic blood pressure is $\bar{x} = 126.07$. Is this evidence that the company's executives in this age group have a different mean systolic blood pressure from the general population?

Solution. To test if the mean is *different* from 128, we use the null hypothesis H_0: $\mu = 128$ with a two-sided alternative H_a: $\mu \neq 128$. Bring up the **Z–Test** screen and adjust the **Inpt** to **STATS**, which allows us to enter the statistics. Enter the values $\mu_0 = 128$, $\sigma = 15$, $\bar{x} = 126.07$, and **n** = 72. Set the alternative to $\neq \mu_0$, then press **ENTER** on either **Calculate** or **Draw**.

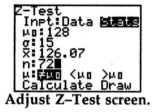
Adjust Z–Test screen.

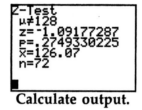
Calculate output.

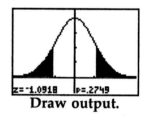
Draw output.

We obtain a z test statistic of -1.0918 and a *P*-value of 0.2749. For this two-sided test, the *P*-value comes from the sum of both tail probabilities: $P(Z \leq -1.0918) + P(Z \geq 1.0918)$. If the true mean for all the company's executives in this age group were equal to 128, then there would be a 27.49% chance of obtaining an $\bar{x}$ as far away as 126.07 with a sample of size 72. This rather high *P*-value does not gives us good evidence to reject the null hypothesis.

Example 6.15 An SRS of 500 California high school seniors gave an average SAT mathematics score of $\bar{x} = 461$. Is this good evidence against the claim that the mean for all California seniors is no more than 450? Assuming that $\sigma = 100$ for all such scores, perform the test H_0: $\mu = 450$, H_a: $\mu > 450$. Give the z test statistic and the *P*-value.

Solution. Bring up the **Z–Test** screen from the **STAT TESTS** menu and adjust the **Inpt** to **STATS**. Enter the value of $\mu_0 = 450$ and the summary statistics, set the alternative to $> \mu_0$, then scroll down to **Calculate** and press **ENTER**.

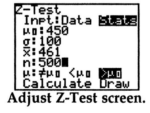
Adjust Z-Test screen.

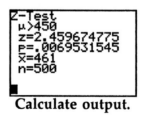
Calculate output.

We obtain a z test statistic of 2.46 and a P-value of 0.00695. Because the P-value is so small, we have significant evidence to reject H_0. For if the true mean were 450, then there would be only a 0.00695 probability of obtaining a sample mean as high as $\bar{x} = 11.2$ with an SRS of 500 students.

Exercise 6.57 The following table gives the DRP scores for a sample of 44 third grade students in a certain district. It is known that $\sigma = 11$ for all such scores in the district. A researcher believes that the mean score of all third graders in this district is higher than the national mean of 32. State the appropriate H_0 and H_a, then conduct the test and give the P-value.

40	26	39	14	42	18	25	43	46	27	19
47	19	26	35	34	15	44	40	38	31	46
52	25	35	35	33	29	34	41	49	28	52
47	35	48	22	33	41	51	27	14	54	45

Solution. Here would should test H_0: $\mu = 32$ with a one-sided alternative H_a: $\mu > 32$. Enter the data into a list, say list **L1**, then call up the **Z–Test** screen and adjust the **Inpt** to **Data**. Enter the values $\mu_0 = 32$ and $\sigma = 11$, set the list to **L1** with frequencies 1, and set the alternative to $> \mu_0$. Then press **ENTER** on **Calculate** or **Draw**.

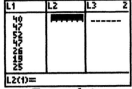
Enter data.

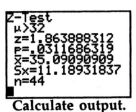
Adjust Z-Test screen.

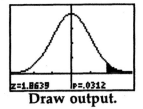
Calculate output.

Draw output.

We obtain a P-value of 0.0311686. If the average of the district were equal to 32, then there would be only a 3.117% chance of a sample group of 44 averaging as high as $\bar{x} = 35.09$. There is strong evidence to reject H_0 and conclude that the district's average is higher than 32.

Exercise 6.71 The readings of 12 radon detectors that were exposed to 105 pCi/l of radon are given below. Assume that $\sigma = 9$ for all detectors exposed to such levels.

91.9	97.8	111.4	122.3	105.4	95.0
103.8	99.6	96.6	119.3	104.8	101.7

(a) Give a 95% confidence interval for the mean reading μ for this type of detector.
(b) Is there significant evidence at the 5% level to conclude that the mean reading differs from the true value of 105? State hypotheses and base a test on the confidence interval from (a).

Solution. (a) We first enter the data into a list, say list **L2**. Then we compute the confidence interval with the **ZInterval** feature from the **STAT TESTS** menu.

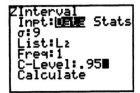
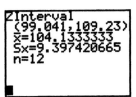

(b) We now test H_0: $\mu = 105$ versus H_a: $\mu \neq 105$. Because the value of 105 falls within the 95% confidence interval (99.041, 109.23) calculated in part (a), we do not have significant evidence at the 5% level to reject the hypothesis that $\mu = 105$.

6.3   Use and Abuse of Tests

We continue with two more exercises that illustrate how one must be careful in drawing conclusions of significance.

Exercise 6.82 Suppose that SATM scores vary normally with $\sigma = 100$. Calculate the P-value for the test of H_0: $\mu = 480$, H_a: $\mu > 480$ in each of the following situations:
(a) A sample of 100 coached students yielded an average of $\bar{x} = 483$.
(b) A sample of 1000 coached students yielded an average of $\bar{x} = 483$.
(c) A sample of 10,000 coached students yielded an average of $\bar{x} = 483$.

Exercise 6.84   For the same hypothesis test as in Exercise 6.82, consider the sample mean of 100 coached students. (a) Is $\bar{x} = 496.4$ significant at the 5% level? (b) Is $\bar{x} = 496.5$ significant at the 5% level?

Solutions. For Exercise 6.82, we adjust the settings in the **Z–Test** screen from the **STAT TESTS** menu and calculate. Below are the results using the three different sample sizes:

 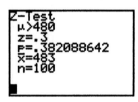 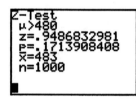 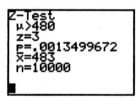

We see that the rise in average to $\bar{x} = 483$ is significant (P very small) only when the results stem from the very large sample of 10,000 coached students. With the sample of only 100 students, there is 38.2% chance of obtaining a sample mean as high as $\bar{x} = 483$, even if the true mean were still 480.

For Exercise 6.84, we perform the **Z–Test** for both values of $\bar{x}$:

 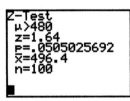 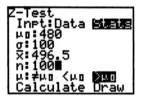 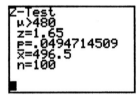

In the first case, $P = 0.0505 > 0.05$; so the value of $\bar{x} = 496.4$ is not significant at the 5% level. However, in the second case, $P = 0.04947 < 0.05$; so the value of $\bar{x} = 496.5$ is significant at the 5% level. However, for SATM scores, there is no real "significant" difference between means of 496.4 and 496.5.

6.4 Power and Inference as a Decision

We conclude this chapter with some exercises on computing the power against an alternative.

Exercise 6.97 Consider the hypotheses H_0: $\mu = 450$, H_a: $\mu > 450$ at the 1% level of significance. A sample of size $n = 500$ is taken from a normal population having $\sigma = 100$. Find the power of this test against the alternative $\mu = 462$.

Solution. We first find the rejection region of the test at the 1% level of significance. Because the alternative is the one-sided right tail, we wish the right-tail probability under the standard normal curve to be 0.01. This probability occurs at $z* = 2.326$. So we reject H_0 if the z test statistic is more than 2.326. That is, we reject if

$$\frac{\bar{x} - 450}{100 / \sqrt{500}} > 2.326$$

or equivalently if $\bar{x} > 450 + 2.326 \times 100 / \sqrt{500} = 460.4022$. Now we must find the probability that $\bar{x}$ is greater than 460.4022, given that the alternative $\mu = 462$ is true.

Given that $\mu = 462$, then $\bar{x} \sim N(462, \ 100 / \sqrt{500})$, and we must compute $P(\bar{x} > 460.4022)$. To do so, we enter the command **normalcdf(460.4022, 1E99, 462, 100/√(500))** and find that the power against the alternative $\mu = 462$ is about 0.64.

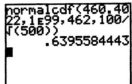

Exercise 6.94 (a) An SRS of size 584 is taken from a population having $\sigma = 58$ to test the hypothesis H_0: $\mu = 100$ versus a two-sided alternative at the 5% level of significance. Find the power against the alternative $\mu = 99$.

Solution. Again, we first must find the rejection regions. For a two-sided alternative at the 5% level of significance, we allow 2.5% at each tail. Thus, we reject if the z test statistic is beyond ± 1.96. That is, we reject if

$$\frac{\bar{x} - 100}{58 / \sqrt{584}} < -1.96 \quad \text{or} \quad \frac{\bar{x} - 100}{58 / \sqrt{584}} > 1.96$$

Equivalently, we reject if $\bar{x} < 95.2959$ or if $\bar{x} > 104.7041$. Now assuming that $\mu = 99$, then $\bar{x} \sim N(99, \ 58 / \sqrt{584})$. We now must compute $P(\bar{x} < 95.2959) + P(\bar{x} > 104.7041)$, which is equivalent to $1 - P(95.2959 \le \bar{x} \le 104.7041) = 1 - $ **normalcdf(95.2959, 104.7041, 99, 58/√(584))**. With this command, we see that the power against the alternative $\mu = 99$ is about 0.07.

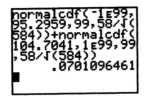
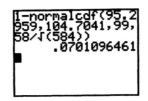

We also could use a program to compute the power against an alternative. In so doing, the results should be more accurate because they will avoid round-off errors in the calculations.

The ZPOWER Program

```
PROGRAM:ZPOWER
:Menu("ZPOWER","ALT. <", 1,
 "ALT. >",2,"ALT. ≠",3,"QUIT",4)
:Lbl 1
:1→C
:Goto 5
:Lbl 2
:2→C
:Goto 5
:Lbl 3
:3→C
:Goto 5
:Lbl 4
:Stop
:Lbl 5
:Disp "TEST MEAN"
:Input M
:Disp "LEVEL OF SIG."
:Input A
:Disp "STANDARD DEV."
:Input S
:Disp "SAMPLE SIZE"
:Input N
```
```
:Disp "ALTERNATIVE"
:Input H
:If C=1
:Then
:invNorm(A,0,1)→Z
:normalcdf(-1E99, M+Z*S/√(N),
 H,S/√(N))→P
:Else
:If C=2
:Then
:invNorm(1-A,0,1)→Z
:normalcdf(M+Z*S/√(N),1E99,
 H,S/√(N))→P
:Else
:invNorm(1-A/2,0,1)→Z
:1-normalcdf(M-Z*S/√(N),
 M+Z*S/√(N),H,S/√(N))→P
:End
:End
:ClrHome
:Disp "POWER"
:Disp P
```

Here are the results of Exercises 6.97 and 6.94 when using the **ZPOWER** program.

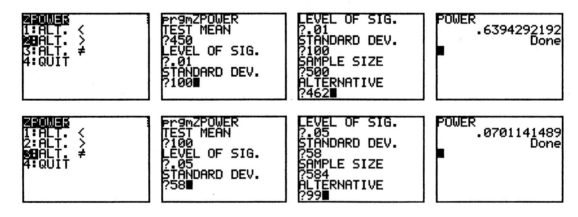

CHAPTER
7

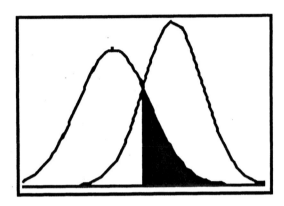

Inference for Distributions

Introduction

In this chapter, we demonstrate the various t procedures that are used for confidence intervals and significance tests about the mean of a normal population for which the standard deviation is unknown.

7.1   Inference for the Mean of a Population

We begin with a short program that allows us to find a critical value $t*$ upon specifying the degrees of freedom and confidence level.

The TSCORE Program

```
PROGRAM:TSCORE
:Disp "DEG. OF FREEDOM"        :"tcdf(0,X,M)"→Y₁
:Input M                        :solve(Y₁-R/2,X,2)→Q
:Disp "CONF. LEVEL"             :Disp "T SCORE"
:Input R                        :Disp round(Q,3)
```

Exercise 7.18 Find the critical values $t*$ for confidence intervals for the mean in the following cases:

(a) A 95% confidence interval based on $n = 20$ observations
(b) A 90% confidence interval from an SRS of 30 observations
(c) An 80% confidence interval from a sample of size 50

Solution. The confidence intervals are based on t distributions with $n - 1$ degrees of freedom. So we need 19 degrees of freedom for part (a), 29 degrees of freedom for part (b), and 49 degrees of freedom for part (c). Below are the outputs of the **TSCORE** program for each part.

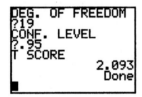 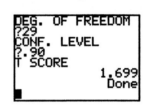 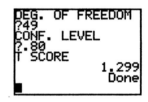

One-sample t Confidence Interval

We now examine confidence intervals for one mean for which we will use the **TInterval** feature (item 8) from the **STAT TESTS** menu. As with the **ZInterval** feature that we used in Chapter 6, we can enter the summary statistics or use data in a list.

Example 7.1 The amount of vitamin C in a factory's production of corn soy blend (CSB) is measured from 8 samples giving $\bar{x} = 22.50$ (mg/100 g) and $s = 7.19$. Find a 95% confidence interval for the mean vitamin C content of the CSB produced during this run.

Solution. Call up the **TInterval** feature from the **STAT TESTS** menu. Set the **Inpt** to **Stats**, then enter the values of $\bar{x}$, **Sx, n, C–Level**, and press **ENTER** on **Calculate**. We obtain the interval 16.489 to 28.511.

 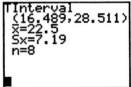

Exercise 7.6 Here are the values of the fuel efficiency in mpg for a sample of trials on a vehicle undergoing testing. Find the mean, the standard deviation, the standard error, the margin of error for a 95% confidence interval, and give a 95% confidence interval for the mean mpg of this vehicle.

15.8	13.6	15.6	19.1	22.4	15.6	22.5	17.2	19.4	22.6
19.4	18.0	14.6	18.7	21.0	14.8	22.6	21.5	14.3	20.9

Solution. First, enter the data into a list, say list **L1**. Next, bring up the **TInterval** screen, set the **Inpt** to **Data**, set the **List** to **L1** with frequencies 1, enter the desired confidence level, and press **ENTER** on **Calculate**. The sample mean, sample deviation, and confidence interval are all displayed.

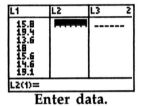

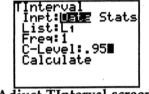

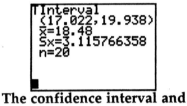

Enter data.   Adjust TInterval screen.   The confidence interval and
                                                                   statistics are displayed.

Because the confidence interval is of the form $\bar{x} \pm m$, we can find the margin of error m by subtracting $\bar{x}$ from the right endpoint of the interval: $m = 19.938 - 18.48 = 1.458$. The standard error is given by $s/\sqrt{n} = 3.115766358/\sqrt{20} \approx 0.6967$.

One-sample t test

We now perform some significance tests about the mean using the **T–Test** feature (item 2) from the **STAT TESTS** menu.

Example 7.3 Using the vitamin C data of $n = 8$, $\bar{x} = 22.50$, and $s = 7.19$ from Example 7.1, test the hypothesis $H_0: \mu = 40$ versus the alternative $H_a: \mu < 40$.

Solution. Bring up the **T–Test** screen from the **STAT TESTS** menu and adjust the **Inpt** to **STATS**. Enter the value of $\mu_0 = 40$ and the summary statistics, set the alternative to $< \mu_0$, then scroll down to **Calculate** and press **ENTER**.

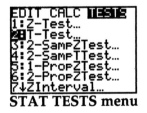

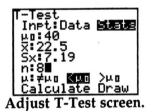

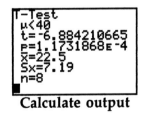

STAT TESTS menu   Adjust T-Test screen.   Calculate output

We obtain a t test statistic of -6.88 and a P-value of 0.0001173. Because the P-value is so small, we have significant evidence to reject H_0. For if the true mean were 40, then there would be only a 0.0001173 probability of obtaining a sample mean as low as $\bar{x} = 22.5$ with a sample of size 8.

Example 7.4 The following table gives the monthly percentage rates of return on a portfolio. Use the data to test the hypothesis $H_0: \mu = 0.95$ versus the alternative $H_a: \mu \neq 0.95$

−8.36	1.63	−2.27	−2.93	−2.70	−2.93	−9.14	−2.64
6.82	−2.35	−3.58	6.13	7.00	−15.25	−8.66	−1.03
−9.16	−1.25	−1.22	−10.27	−5.11	−0.80	−1.44	1.28
−0.65	4.34	12.22	−7.21	−0.09	7.34	5.04	−7.24
−2.14	−1.01	−1.41	12.03	−2.56	4.33	2.35	

Solution. Enter the data into a list, say list **L2**, then bring up the **T–Test** screen and adjust the **Inpt** to **Data**. Enter the value of $\mu_0 = .95$, set the list to **L2** with frequencies 1, and set the alternative to $\neq \mu_0$. Then press **ENTER** on **Calculate** or **Draw**.

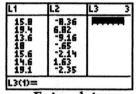
Enter data.

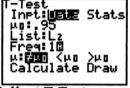
Adjust T–Test screen.

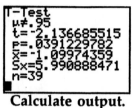
Calculate output.

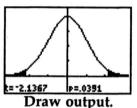
Draw output.

We obtain a *P*-value of 0.0391. If the average return were equal to 0.95%, then there would be only a 3.91% chance of a sample of 39 months averaging as far way as $\bar{x} = -1.1\%$. There is sufficient evidence to reject H_0 and conclude that the mean monthly return differs from 0.95%.

Matched Pair *t* Procedure

Exercise 7.27 Two operators of X-ray machinery measured the same eight subjects for total body bone mineral content. Here are the results in grams:

	Subject							
Operator	1	2	3	4	5	6	7	8
1	1.328	1.342	1.075	1.228	0.939	1.004	1.178	1.286
2	1.323	1.322	1.073	1.233	0.934	1.019	1.184	1.304

Use a significance test to examine the null hypothesis that the two operators have the same mean. Use a 95% a confidence interval to provide a range of differences that are compatible with these data.

Solution. We consider the average μ_D of the *difference* of the measurements between the operators. First, enter the measurements of Operator 1 into list **L1** and the measurements of Operator 2 into list **L2**. Next, use the command **L1–L2→L3** to enter the differences into list **L3**. Then, use a **T–Test** on list **L3** to test $H_0: \mu_D = 0$ versus the alternative $H_a: \mu_D \neq 0$.

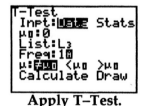
Enter data.   Store differences.   Apply T–Test.   Calculate output.

We obtain a *P*-value of 0.7387 from a test statistic of $t = -0.347$. Due to the high *P*-value, we can say that there is not a significant average difference. For if μ_D were equal to 0, then there would be a 73.87% chance of having an average difference as far away as $\bar{d} = -0.0015$ with a random sample of 8 subjects.

Next, use the **TInterval** on list **L3** to find a 95% confidence interval for the average difference. Because the interval (–0.0117, 0.00872) contains 0, we have further evidence that the operators could have the same mean.

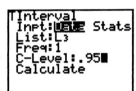 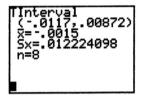

The Power of the *t* test

As in Section 6.4, we also can compute the power of a *t* test against an alternative. Below we work an example "by hand" and with a program.

Example 7.9 Consider the hypothesis test $H_0: \mu = 0$, $H_a: \mu > 0$, with $\alpha = 0.05$. Taking $s = 2.5$ and $\sigma = 3$, find the power against the alternative $\mu = 1.6$ for a sample of size $n = 20$.

Solution. We first find the rejection region of the test at the 5% level of significance. For this one-sided alternative, we have 5% probability at the right tail, which corresponds to the critical values t^* of a 90% confidence interval using the $t(n - 1) = t(19)$ distribution.

Using the **TSCORE** program, we find that the t^* value is 1.729. Thus, we reject H_0 if $\dfrac{\bar{x} - 0}{2.5 / \sqrt{20}} > 1.729$, which means that we reject H_0 if $\bar{x} > 0.96654$.

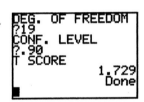

We now must compute the probability that $\bar{x}$ falls in this rejection region, given that the alternative $\mu = 1.6$ is true. Now given that $\mu = 1.6$ and $\sigma = 3$, then $\bar{x} \sim N(1.6, 3 / \sqrt{20}) = N(1.6, 0.67082)$. Using this distribution, we must compute $P(\bar{x} > 0.96654)$.

For this calculation, we use the built-in **normalcdf(** command from the **DISTR** menu and enter **normalcdf(.96654, 1E99, 1.6, .67082)**. We see that the power against the alternative $\mu = 1.6$ is 0.8275. If $\mu = 1.6$, and $\sigma = 3$, then we are 82.75% likely to reject H_0.

Below are the results from using the **TPOWER** program provided on the next page.

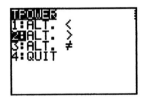 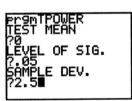  

The TPOWER Program

```
PROGRAM:TPOWER                          :Input N
:Menu("TPOWER","ALT.<",1,               :Disp "ALTERNATIVE"
  "ALT. >",2,"ALT. ≠",3,"QUIT",4)       :Input H
:Lbl 1                                  :"tcdf(0,H,N-1)"→Y₁
:1→C                                    :If C=1
:Goto 5                                 :Then
:Lbl 2                                  :solve(Y₁-(0.5-A),H,2)→Q
:2→C                                    :normalcdf(-1E99,M-Q*S/√(N),
:Goto 5                                   H,T/√(N))→P
:Lbl 3                                  :Else
:3→C                                    :If C=2
:Goto 5                                 :Then
:Lbl 4                                  :solve(Y₁-(0.5-A),H,2)→Q
:Stop                                   :normalcdf(M+Q*S/√(N),1E99,
:Lbl 5                                     H,T/√(N))→P
:Disp "TEST MEAN"                       :Else
:Input M                                :solve(Y₁-(0.5-A/2),H,2)→Q
:Disp "LEVEL OF SIG."                   :1-normalcdf(M-Q*S/√(N),
:Input A                                  M+Q*S/√(N),H,T/√(N))→P
:Disp "SAMPLE DEV."                     :End
:Input S                                :End
:Disp "TRUE ST. DEV."                   :ClrHome
:Input T                                :Disp "POWER"
:Disp "SAMPLE SIZE"                     :Disp P
```

The Sign Test

Example 7.12 Out of 15 patients, 14 had more aggressive behavior on moon days than on other days. Use the sign test on the hypothesis of "no moon effect."

Solution. Because so many patients had a change in behavior, we shall test the hypothesis H_0: $p = 0.50$ with the alternative H_a: $p > 0.50$.

We must compute the probability of there being as many as 14 changes with a $B(15, 0.50)$ distribution. Equivalently, we can compute the probability of there being as few as one non-change. Thus, we could compute either $P(B \geq 14) = 1 - P(B \leq 13)$ or $P(B \leq 1)$. To do so, we use the built-in **binomcdf(** command from the **DISTR** menu.

After entering either **1− binomcdf(15, .5, 13)** or **binomcdf(15, .5, 1)**, we obtain the very low P-value of 0.000488. If there were no moon effect, then there would be almost no chance of having as many as 14 out of 15 showing a change. Therefore, we can reject H_0 in favor of the alternative that the moon generally causes more aggressive behavior.

```
1-binomcdf(15,.5
,13)
       4.8828125E-4
binomcdf(15,.5,1
)
       4.8828125E-4
```

7.2 Comparing Two Means

We next consider confidence intervals and significance tests for the difference of means $\mu_1 - \mu_2$ given two normal populations that have unknown standard deviations. The results are based on independent random samples of sizes n_1 and n_2. For the most accurate results, we can use the **2–SampTInt** and **2–SampTTest** features from the **STAT TESTS** menu.

These features require that we specify whether or not we wish to use the pooled sample variance s_p^2. We should specify "Yes" only when we assume that the two populations have the *same* (unknown) variance. In this case, the critical values $t*$ are obtained from the $t(n_1 + n_2 - 2)$ distribution and the standard error is $s_p\sqrt{1/n_1 + 1/n_2}$. When we specify "No" for the pooled variance, then the standard error is $\sqrt{s_1^2/n_1 + s_2^2/n_2}$ and the degrees of freedom r are given by

$$r = \frac{\left(\dfrac{s_1^2}{n_1} + \dfrac{s_2^2}{n_2}\right)^2}{\dfrac{1}{n_1-1}\left(\dfrac{s_1^2}{n_1}\right)^2 + \dfrac{1}{n_2-1}\left(\dfrac{s_2^2}{n_2}\right)^2}$$

But if the true population standard deviations σ_1 and σ_2 are known, then we should use the **2–SampZInt** and **2–SampZTest** features for our calculations.

We can also use the **TWOTCI and TWOTTEST** programs that follow in order to calculate the less accurate results, where the critical value $t*$ is obtained from the t distribution having degrees of freedom that is the smaller of $n_1 - 1$ and $n_2 - 1$.

The TWOTCI Program

```
Program:TWOTCI              :Disp "CONF. LEVEL"
:Disp "X SAMPLE SIZE"       :Input R
:Input P                    :min(P-1,Q-1)→N
:Disp "XBAR"                :"tcdf(0,X,N)"→Y₁
:Input X                    :solve(Y₁-R/2,X,2)→B
:Disp "X SAMPLE DEV."       :B√(S²/P+T²/Q)→E
:Input S                    :ClrHome
:Disp "Y SAMPLE SIZE"       :Disp "DIFF, ERROR"
:Input Q                    :Disp round(X-Y,4)
:Disp "YBAR"                :Disp round(E,4)
:Input Y                    :Disp "INTERVAL"
:Disp "Y SAMPLE DEV."       :Disp round(X-Y-E,4)
:Input T                    :Disp round(X-Y+E,4)
```

The TWOTTEST Program

```
Program:TWOTTEST                    :Input T
:Menu("TWOTTEST","ALT. <",1,"ALT. >",2,    :(X-Y)/√(S²/P+T²/Q)→Z
  "ALT. ≠",3,"QUIT",4)              :min(P-1,Q-1)→N
:Lbl 1                              :If Z≥0
:1→C                                :Then
:Goto 5                             :0.5-tcdf(0,Z,N)→R
:Lbl 2                              :1-R→L
:2→C                                :Else
:Goto 5                             :0.5-tcdf(Z,0,N)→L
:Lbl 3                              :1-L→R
:3→C                                :End
:Goto 5                             :ClrHome
:Lbl 4                              :Disp "T STAT"
:Stop                               :Disp Z
:Lbl 5                              :Disp "P VALUE"
:Disp "X SAMPLE SIZE"               :If C=1
:Input P                            :Then
:Disp "XBAR"                        :Disp L
:Input X                            :Else
:Disp "X SAMPLE DEV."               :If C=2
:Input S                            :Then
:Disp "Y SAMPLE SIZE"               :Disp R
:Input Q                            :Else
:Disp "YBAR"                        :Disp 2*min(L,R)
:Input Y                            :End
:Disp "Y SAMPLE DEV."               :End
```

Examples 7.14, 7.15 Two groups of students were given a DRP test. The results are given in the table below. Test the hypothesis H_0: $\mu_1 = \mu_2$ versus H_a: $\mu_1 > \mu_2$. Give a 95% confidence interval for $\mu_1 - \mu_2$.

Group	n	$\bar{x}$	s
Treatment	21	51.48	11.01
Control	23	41.52	17.15

Solution. Bring up the **2–SampTTest** feature from the **STAT TESTS** menu, and set the **Inpt** to **Stats.** Enter the given statistics, set the alternative, and enter **No** for **Pooled.** Then press **ENTER** on **Calculate.**

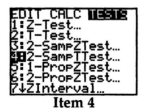
Item 4

Enter data.

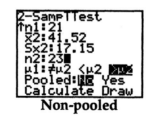
Non-pooled

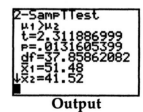
Output

We obtain a *P*-value of 0.01316 from a test statistic of 2.3119 with 37.85862 degrees of freedom. If the true means were equal, then there would be a very small chance of $\bar{x}_1$ being so much larger than $\bar{x}_2$ with samples of these sizes. We therefore can reject H_0 and conclude that $\mu_1 > \mu_2$.

Alternately, we can use the **TWOTTEST** program that uses the *t* distribution having degrees of freedom that is the smaller of $n_1 - 1$ and $n_2 - 1$. The results are shown below.

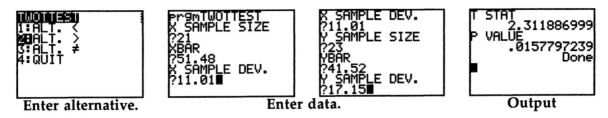

| Enter alternative. | Enter data. | | Output |

To calculate a confidence interval for $\mu_1 - \mu_2$, bring up the **2–SampTInt** screen (item 0 in the **STAT TESTS** menu), set the **Inpt** to **Stats**, enter the given statistics and desired confidence level, and calculate. We obtain the interval (1.2375, 18.683).

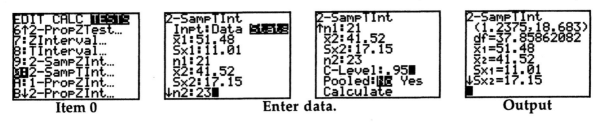

| Item 0 | Enter data. | | Output |

Below is the confidence interval obtained by executing the **TWOTCI** program that uses the $t*$ value with the degrees of freedom being the smaller of $n_1 - 1$ and $n_2 - 1$.

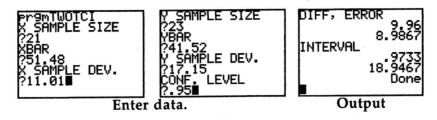

| Enter data. | Output |

Exercise 7.84. The Survey of Study Habits and Attitudes was given to first-year students at a private college. The tables below show a random sample of the scores.

Women's scores

154	109	137	115	152	140	154	178	101
103	126	126	137	165	165	129	200	148

Men's scores

108	140	114	91	180	115	126	92	169	146
109	132	75	88	113	151	70	115	187	104

(a) Examine each sample graphically to determine if the use of a t procedure is acceptable.

(b) Test the supposition that the mean score for all men is lower than the mean score for all women among first-year students at this college.

(c) Give a 90% confidence interval for the mean difference between the SSHA scores of male and female first-year students at this college.

Solution. (a) We shall make normal quantile plots of these data. In the **STAT EDIT** screen, enter the women's scores into list **L1** and the men's scores into list **L2**. Choose an appropriate **WINDOW** with an **X** range that allows you to see the minimum and maximum of both data sets and a **Y** range from –3 to 3.

In the **STAT PLOT** screen, adjust the **Type** settings for both **Plot1** and **Plot2** to the sixth type for normal quantile plot, then graph each plot separately. The resulting plots appear close enough to linear to warrant use of t procedures.

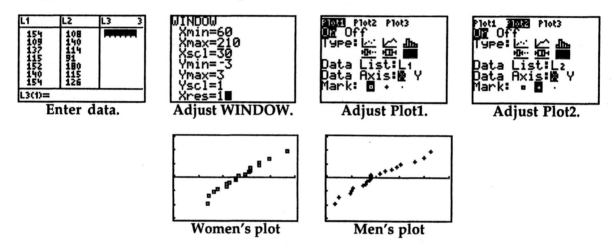

Enter data.   Adjust WINDOW.   Adjust Plot1.   Adjust Plot2.

Women's plot   Men's plot

(b) Next, let μ_1 be the mean SSHA score among all first-year women and let μ_2 be the mean score among all first-year men. We shall test the hypothesis $H_0: \mu_1 = \mu_2$ versus the alternative $H_a: \mu_1 > \mu_2$. In the **2–SampTTest** screen, set the **Inpt** to **Data**, enter the desired lists **L1** and **L2,** set the alternative to $> \mu_2$, enter **No** for **Pooled**, and press **ENTER** on **Calculate** or **Draw**.

 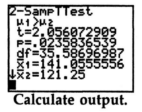 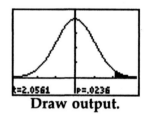

Adjust 2–SampTTest.   Calculate output.   Draw output.

We obtain a P-value of 0.02358. If the true means were equal, then there would be only a 2.358% chance of $\bar{x}_1$ being so much larger than $\bar{x}_2$ with samples of these sizes. The relatively low P-value gives us evidence to reject H_0 and conclude that $\mu_1 > \mu_2$. That is, the mean score for all men is lower than the mean score for all women among first-year students at this college.

(c) Adjust the settings in the **2–SampTInt** screen and calculate. We obtain (3.5377, 36.073). That is, the mean score of female first-year students should be from about 3.5377 points higher to about 36.073 points higher than the mean score of male first-year students at this college.

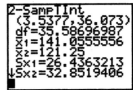

Pooled Two-sample *t* Procedures

Exercise 7.83 Here are the summary results on hemoglobin levels at 12 months of age for two samples of infants:

Group	n	$\bar{x}$	s
Breast-fed	23	13.3	1.7
Formula	19	12.4	1.8

(a) Is there significant evidence that the mean hemoglobin level is higher among breast-fed babies? State H_0 and H_a, and carry out a *t* test.
(b) Give a 95% confidence interval for the difference in mean hemoglobin levels between the two populations of infants.

Solution. Let μ_1 be the mean hemoglobin level for all breast-fed babies and let μ_2 be the mean level for all formula-fed babies. Because the sample deviations are so close, it appears that true standard deviations among the two groups could be equal; thus, we may use the pooled two-sample *t* procedures. For part (a), we will test H_0: $\mu_1 = \mu_2$ versus H_a: $\mu_1 > \mu_2$.

Call up the **2–SampTTest** feature, and set the **Inpt** to **Stats**. Enter the given statistics, set the alternative, enter **Yes** for **Pooled**, and calculate.

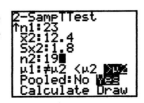

With a P-value of 0.052, we conclude that there is not statistical evidence (at the 5% level) to reject H_0. If the true means were equal, then there is greater than a 5% chance of $\bar{x}_1$ being 0.9 higher than $\bar{x}_2$ with samples of these sizes.

(b) Next, calculate a 95% confidence interval with the **2–SampTInt** feature set to **Yes** on **Pooled**. We see that $-0.1938 \leq \mu_1 - \mu_2 \leq 1.9938$. That is, the mean level of breast-fed babies could be from 0.1938 lower to 1.9938 higher than the mean level of formula-fed babies.

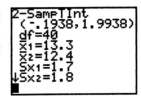

7.3 Optional Topics in Comparing Distributions

We now demonstrate a test for determining whether or not two normal populations have the same variance. If so, then we would be justified in using the pooled two-sample *t* procedures for confidence intervals and significance tests about the difference in means. For the test, we will need the **2–SampFTest** feature (item D) from the **STAT TESTS** menu.

The *F* Ratio Test

Exercise 7.101 Consider again the data from Exercise 7.84 regarding the SSHA scores of first-year students at a private college. Test whether the women's scores are less variable.

Women's scores

154	109	137	115	152	140	154	178	101
103	126	126	137	165	165	129	200	148

Men's scores

108	140	114	91	180	115	126	92	169	146
109	132	75	88	113	151	70	115	187	104

Solution. Let σ_1 be the standard deviation of all women's scores and let σ_2 be the standard deviation for all men's scores. We shall test $H_0: \sigma_1 = \sigma_2$ versus $H_a: \sigma_1 < \sigma_2$. To do so, first enter the data sets into lists, say **L1** and **L2**. Next, bring up the **2–SampFTest** screen from the **STAT TESTS** menu, set the **Inpt** to **Data**, enter the appropriate lists and alternative, and calculate.

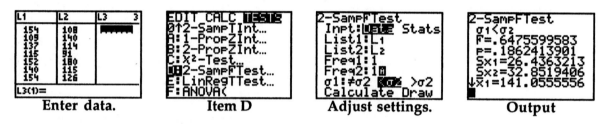

Enter data.   Item D   Adjust settings.   Output

With a *P*-value of 0.18624, we do not have strong evidence to reject H_0. If σ_1 were equal to σ_2, then there would be an 18.6% chance of the women's sample deviation of **Sx1** = 26.4363 being so much lower than the men's sample deviation of **Sx2** = 32.8519. Thus, we cannot assert strongly that the women's scores are less variable.

Example 7.22 Here are the summary statistics for drop in blood pressure among two sample groups of patients undergoing treatment. Test to see if the groups in general have the same standard deviation.

Group	n	$\bar{x}$	s
Calcium	10	5.000	8.743
Placebo	11	−0.273	5.901

Solution. Let σ_1 be the standard deviation of all possible patients in the calcium group, and let σ_2 be the standard deviation of all possible patients in the placebo group. We will test the hypothesis $H_0: \sigma_1 = \sigma_2$ versus $H_a: \sigma_1 \neq \sigma_2$. Bring up the **2–SampFTest** screen and set the **Inpt** to **Stats**. Enter the summary statistics and alternative, then calculate.

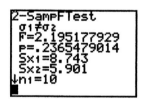

We obtain an *F*-statistic of 2.195 and a *P*-value of 0.2365479. If σ_1 were equal to σ_2, then there would be about a 23.6% chance of **Sx1** and **Sx2** being so far apart with samples of these sizes. This evidence may not be significant enough to reject H_0 in favor of the alternative.

The Power of the Two-Sample *t* test

We conclude this chapter with a program that gives a standard normal approximation of the power of the pooled two-sample *t* test.

The POWER2T Program

```
PROGRAM:POWER2T          :Input S
:Disp "ALT. MEAN DIFF."  :0.5-A→R
:Input L                 :N+M-2→F
:Disp "1ST SAMPLE SIZE"  :"tcdf(0,X,F)"→Y₁
:Input N                 :solve(Y₁-R,X,2)→Q
:Disp "2ND SAMPLE SIZE"  :abs(L)/S/√(1/N+1/M)→D
:Input M                 :0.5-normalcdf(0,Q-D,0,1)→P
:Disp "LEVEL OF SIG."    :ClrHome
:Input A                 :Disp "POWER"
:Disp "COMMON ST. DEV"   :Disp round(P,4)
```

Example 7.23 Find a normal approximation of the power of the two-sample *t* test with the following design: An alternative difference of $\mu_1 - \mu_2 = 5$, samples of sizes $n_1 = n_2 = 45$, a level of significance of $\alpha = 0.01$, and an assumed common standard deviation of 7.4.

Solution. Simply execute the **POWER2T** program to obtain the normal approximation of the power as 0.7983.

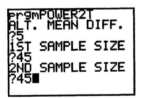 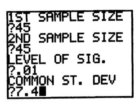 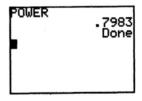

CHAPTER

8

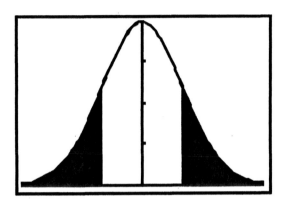

Inference for Proportions

8.1 Inference for a Single Proportion
8.2 Comparing Two Proportions

Introduction

In this chapter, we discuss how to use the TI-83 Plus to find confidence intervals and to conduct hypothesis tests for a single proportion and for the difference in two population proportions. The confidence intervals will be computed both with provided programs and with the built-in **1–PropZInt** and **2–PropZInt** features.

8.1 Inference for a Single Proportion

Both the large-sample and plus-four level C confidence intervals can be calculated using the **1–PropZInt** feature (item A) from the **STAT TESTS** menu. Significance tests can be worked using the **1–PropZTest** (item 5).

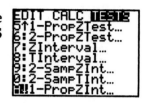

Large-Sample Confidence Interval

Example 8.5 In restaurant worker survey, 68 of a sample of 100 employees agreed that work stress had a negative impact on their personal lives. Find a 95% confidence interval for the true proportion of restaurant employees who agree.

Solution. Bring up the **1–PropZInt** screen, enter **68** for **x**, enter **100** for **n**, and enter **.95** for **C–Level**. Then press enter on **Calculate** to obtain a 95% confidence interval of (0.58857, 0.77143).

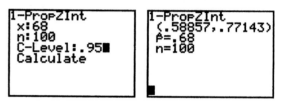

Plus-Four Confidence Interval

Example 8.2 In a preliminary sample of 12 female subjects, it was found that 4 were equol producers. Find a 95% confidence interval for the true proportion of females who are equol producers.

Solution. In the **1–PropZInt** screen, enter **6** for **x**, which is 2 more than the actual number of "Yes," and enter **26** for **n**, which is 4 more than the actual sample size. Enter the desired **C–level** and press **ENTER** on **Calculate** to obtain the plus-four estimate $\hat{p} = 0.375$ and the confidence interval.

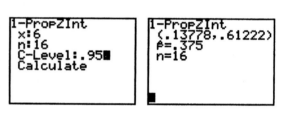

Choosing a Sample Size

As with confidence intervals for the mean, we often would like to know in advance what sample size would provide a certain maximum margin of error m with a certain level of confidence. The required sample size n satisfies

$$n \geq \left(\frac{z*}{m}\right)^2 p*(1 - p*)$$

where $z*$ is the appropriate critical value depending on the level of confidence and $p*$ is a guessed value of the true proportion p. If $p* = 0.50$, then the resulting sample size insures that the margin of error is no more than m, regardless of the true value of p.

The program **PSAMPSZE** displays the required sample size, rounded up to the nearest integer, after one enters the desired error m, the confidence level, and the guess $p*$.

The PSAMPSZE Program

PROGRAM:PSAMPSZE	**:If int(M)=M**
:Disp "DESIRED ERROR"	**:Then**
:Input E	**:M→N**
:Disp "CONF. LEVEL"	**:Else**
:Input R	**:int(M+1)→N**
:Disp "GUESS OF P"	**:End**
:Input P	**:ClrHome**
:invNorm((R+1)/2,0,1)→Q	**:Disp "SAMPLE SIZE="**
:(Q/E)²*P(1-P)→M	**:Disp int(N)**

Exercise 8.26 Among students who completed an alcohol awareness program, you want to estimate the proportion who state that their behavior towards alcohol has changed since the program. Using the guessed value of $p* = 0.30$ from previous surveys, find the sample size required to obtain a 95% confidence interval with a maximum margin of error of $m = 0.10$.

Solution. Executing the **PSAMPSZE** program, we find that an sample of size 81 would be required. This value also can be obtained by

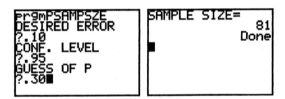

$$n = (1.96 / 0.10)^2 \times 0.30 \times 0.70 = 80.6736$$

Significance Tests

We now show how to conduct hypothesis tests for a single population proportion p using the **1–PropZTest** feature (item 5) in the **STAT TESTS** menu.

Example 8.3 In the restaurant worker survey, 68 of a sample of 100 employees agreed that work stress had a negative impact on their personal lives. Let p be the true proportion of restaurant employees who agree. Test the hypothesis $H_0: p = 0.75$ versus $H_a: p \neq 0.75$.

Solution. Enter the data and alternative into the **1–PropZTest** screen, then press enter on **Calculate** or **Draw**. We obtain a (two-sided) P-value of 0.106 from a z-statistic of -1.61658. If p were equal to 0.75, then there would be a 10.6% chance of obtaining $\hat{p}$ as far away as 0.68 with a sample of size 100.

Item 5

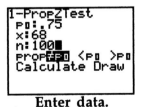
Enter data.

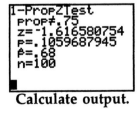
Calculate output.

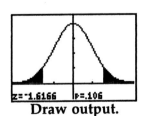
Draw output.

Exercise 8.22 In a taste test of instant versus fresh-brewed coffee, only 12 out of 40 subjects preferred the instant coffee. Let p be the true probability that a random person prefers the instant coffee. Test the claim H_0: $p = 0.50$ versus H_a: $p < 0.50$ at the 5% level of significance.

Solution. Enter the data and alternative into the **1–PropZTest** screen and calculate. We obtain a test statistic of –2.53 and a P-value of 0.0057. If p were 0.50, then there would be only a 0.0057 probability of $\hat{p}$ being as low as 0.3 with 40 subjects. There is strong evidence to reject H_0.

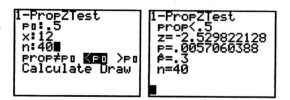

8.2   Comparing Two Proportions

We now demonstrate confidence intervals and significance tests for the difference of two population proportions p_1 and p_2. These calculations can be made with the **2–PropZInt** (item B) and **2–PropZTest** (item 5) features from the **STAT TESTS** menu.

Large-Sample Confidence Interval for Difference of Proportions

Example 8.9 The table below gives the sample sizes and numbers of men and women who responded "Yes" to being frequent binge drinkers in a survey of college students. Find a 95% confidence interval for the difference between the proportions of men and women who are frequent binge drinkers.

Population	n	X
Men	7180	1630
Women	9916	1684

Solution. In the **2–PropZInt** screen, enter **1630** for **x1**, **7180** for **n1**, **1684** for **x2**, and **9916** for **n2**. Set the confidence level to **.95** and press **ENTER** on **Calculate**.

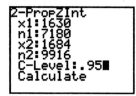 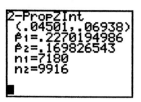

We obtain a confidence interval of (0.045, 0.069). That is, the proportion of male binge drinkers is from 4.5 percentage points higher to 6.9 percentage points higher than the proportion of female binge drinkers.

Plus–Four Confidence Interval for Difference of Proportions

Example 8.10 A study of 12 boys and 12 girls found that four of the boys and three of the girls had a Tanner score of 4 or 5. Find a plus-four 95% confidence interval for the difference in proportions of all boys and all girls who would score a 4 or 5.

Solution. We can still use the **2–PropZInt** feature to find a plus-four confidence interval for $p_1 - p_2$. But for **x1** and **x2**, enter 1 more than the actual number of positive responses. For **n1** and **n2**, enter 2 more than the actual sample sizes. Here, enter **5** for **x1**, **4** for **x2**, and enter **14** for both **n1** and **n2**. Then, set the confidence level to .95 and calculate.

We see that $-0.2735 \le p_1 - p_2 \le 0.4164$. That is, the true proportion of boys who score 4 or 5 is from 27.35 percentage points lower to 41.64 percentage points higher than the true proportion of girls who score 4 or 5.

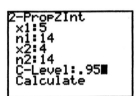 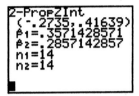

Significance Tests for Difference of Proportions

We now show how to conduct hypothesis tests about $p_1 - p_2$ using the **2–PropZTest** feature (item 6) from the **STAT TESTS** menu.

Example 8.11 The table below gives the sample sizes and numbers of men and women who responded "Yes" to being frequent binge drinkers in a survey of college students. Does the data give good evidence that the true proportions of male binge drinkers and female binge drinkers are different?

Population	n	X
Men	7180	1630
Women	9916	1684

Solution. Let p_1 be the true proportion of male students who are frequent binge drinkers, and let p_2 be the true proportion of female students who are frequent binge drinkers. We shall test the hypothesis $H_0: p_1 = p_2$ versus the alternative $H_a: p_1 \ne p_2$. Bring up the **2–PropZTest** screen, enter the actual data and the two-sided alternative, then calculate.

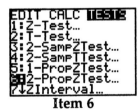  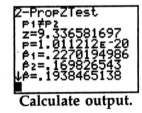

Item 6	Enter data.	Calculate output.

We obtain a P-value of 1.0112×10^{-20} (essentially 0) from a test statistic of $z = 9.33658$. If $p_1 = p_2$ were true, then there would be no chance of obtaining sample proportions as far apart as $\hat{p}_1 = 0.227$ and $\hat{p}_2 = 0.1698$ with samples of these sizes. So, we can reject H_0.

Exercise 8.51 The table below gives the results of a gender bias analysis of a textbook. Do the data give evidence that the proportion of juvenile female references is higher than the proportion of juvenile male references? Compare the results with those of a 90% confidence interval.

Gender	n	X (juvenile)
Female	60	48
Male	132	52

Solution. Let p_1 be the true proportion of juvenile female references in all such texts, and let p_2 be the true proportion of juvenile male references. We will test the hypothesis H_0: $p_1 = p_2$ versus H_a: $p_1 > p_2$. Bring up the **2–PropZTest** screen, enter the data, set the alternative to **>p2**, and calculate. Then bring up the **2–PropZInt** screen, and calculate a 90% confidence interval.

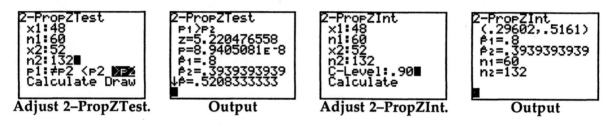

Adjust 2–PropZTest.   **Output**   **Adjust 2–PropZInt.**   **Output**

We obtain a very low P-value of 8.94×10^{-8}, which gives strong evidence to reject H_0. If $p_1 = p_2$ were true, then there would be almost no chance of obtaining a $\hat{p}_1$ that is so much higher than $\hat{p}_2$ with samples of these sizes.

The 90% confidence interval states that $0.296 \le p_1 - p_2 \le 0.5161$. That is, among all such texts, the proportion of juvenile female references is from 29.6 percentage points higher to 51.61 percentage points higher than the proportion of juvenile male references.

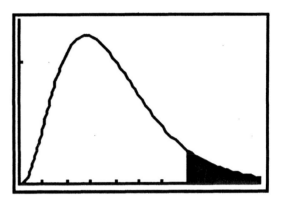

Inference for Two-Way Tables

Introduction

In this chapter, we describe how to use the TI-83 Plus to perform a chi-square test on data from a two-way table. We shall be testing whether there is any association between the row variable traits and the column variable traits, or whether these row and column traits are independent.

9.1   Data Analysis for Two-Way Tables

We first provide a program that converts a two-way table of raw data into three different proportion tables.

The TWOWAY Program

```
PROGRAM:TWOWAY                          :1→[B](R+1,C+1)
:Disp "NO. OF ROWS"                     :{R+1,C}→dim([C])
:Input R                                :For(J,1,C)
:Disp "NO. OF COLUMNS"                  :sum(seq([A](I,J),I,1,R))→A
:Input C                                :For(I,1,R)
:{R+1,C+1}→dim([B])                      :round([A](I,J)/A,4)→[C](I,J)
:0→N                                    :End
:For(I,1,R)                             :1→[C](R+1,J)
:N+sum(seq([A](I,J),J,1,C))→N           :End
:End                                    :{R,C+1}→dim([D])
:For(I,1,R)                             :For(I,1,R)
:For(J,1,C)                             :sum(seq([A](I,J),J,1,C))→A
:round([A](I,J)/N,4)→[B](I,J)            :For(J,1,C)
:End                                    :round([A](I,J)/A,4)→[D](I,J)
:End                                    :End
:For(I,1,R)                             :1→[D](I,C+1)
:round(sum(seq([A](I,J),J,1,C))         :End
 /N,4)→[B](I,C+1)                        :ClrHome
:End                                    :Output(2,1,"JT/MARG: MAT B")
:For(J,1,C)                             :Output(4,1,"COND.GIVEN THE
:round(sum(seq([A](I,J),I,1,R))          COL VAR: MAT C")
 /N,4)→[B](R+1,J)                        :Output(7,1,"COND.GIVEN THE
:End                                     ROW VAR: MAT D")
```

Before executing the program, enter the raw data (excluding totals) into matrix **[A]** in the **MATRX EDIT** screen. The program then stores proportion tables into matrices **[B]**, **[C]**, and **[D]**. Matrix **[B]** gives the joint distribution of overall proportions from the entire sample as well as the marginal distributions; matrix **[C]** gives the conditional distribution given the column variable, and matrix **[D]** gives the conditional distribution given the row variable.

Exercise 9.9 Questionnaires were mailed to 300 randomly selected businesses in each of three categorical sizes. The following data show the number of responses.

Size of company	Response	No response	Total
Small	175	125	300
Medium	145	155	300
Large	120	180	300

(a) What was the overall percent of non-response?
(b) For each size of company, compute the non-response percentage.
(c) Draw a bar graph of the non-response percents.
(d) Using the total number of responses as a base, compute the percentage of responses that came from each size of company.

Solution. We can compute all three types of percentages with the **TWOWAY** program. Not including the totals, the data creates a 3 × 2 matrix. There are 3 rows (sizes of companies) and 2 columns (Response and No response). Before running the program, we must enter this data into matrix [A] in the **MATRX EDIT** screen. After entering the data, execute the **TWOWAY** program by entering the dimensions of 3 rows and 2 columns when prompted.

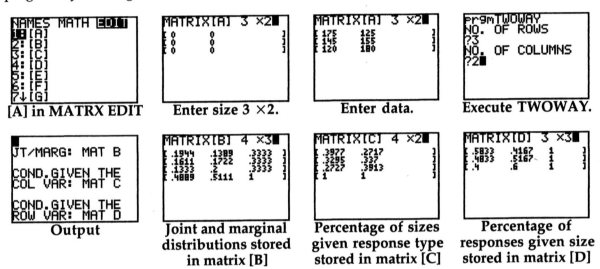

| [A] in MATRX EDIT | Enter size 3 ×2. | Enter data. | Execute TWOWAY. |

| Output | Joint and marginal distributions stored in matrix [B] | Percentage of sizes given response type stored in matrix [C] | Percentage of responses given size stored in matrix [D] |

(a) Matrix [B] contains the marginal distribution of response/non-response rates. From the last row of matrix [B], we see that of the 900 questionnaires, 48.89% responded and 51.11% did not respond. From the last column, we see that one-third of the questionnaires went to each of the small, medium, and large companies, which is the marginal distribution of company sizes.

The inner entries of matrix [B] contain the joint distribution. For instance, from the second column we see that among all 900 questionnaires, 13.89% were small company non-responses, 17.22% were medium company non-responses, and 20% were large company non-responses.

(b) Matrix [D] contains the conditional distribution of responses given type of company size. From the second column in matrix [D], we see that 41.67% of small companies did not respond, which is 125 out of 300. Among medium-sized companies, 51.67% did not respond, which is 155 out of 300. Among large companies, 60% did not respond, which is 180 out of 300.

(d) Matrix [C] contains the conditional distribution of company size given type of response. From the first column in matrix [C], we see that among those that responded, 39.77% were small companies (175 out of 440), 32.95% were medium-sized companies (145 out of 440), and 27.27% were large companies (120 out of 440).

(c) To make a bar graph of the non-response percents, we first enter the values **0, 1**, and **2** into list **L1** in order to represent the three types of companies, and then enter the non-response proportions from the second column of matrix [D] into list **L2**. Next, we adjust the **WINDOW** and **STAT PLOT** settings for a histogram of **L1** with frequencies **L2**, then graph.

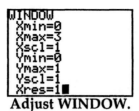

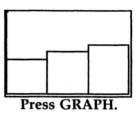

Enter three categories and non-response percents.　Adjust WINDOW.　Adjust STAT PLOT.　Press GRAPH.

9.2 Inference for Two-Way Tables

We continue here with an example that shows how to compute the expected cell counts of a two-way table under the hypothesis that there is no association between the row variable and the column variable.

Expected Cell Counts

The TI-83 Plus has a built-in χ^2–Test feature (item C in the **STAT TESTS** menu) that will compute the expected counts of a random sample under the assumption that the conditional distributions are the same for each category type. To use this feature, we first must enter data from a two-way table into a matrix in the **MATRX EDIT** screen.

Example 9.12 The following table shows the two-way relationship between whether a franchise succeeds and whether it has exclusive territory rights for a number of businesses.

Observed number of firms			
	Exclusive territory		
Success	Yes	No	Total
Yes	108	15	123
No	34	13	47
Total	142	28	170

Under the assumption that there is no relationship between success and exclusive territory rights, find the expected number of successful franchises for each type of firm.

Solution. First, enter the 2×2 table of data (excluding totals) into matrix [A] in the **MATRX EDIT** screen. Next, bring up the χ^2–Test screen from the **STAT TESTS** menu, and adjust the **Observed** and **Expected** settings. To enter matrix [A] for **Observed**, press **MATRX** and then press **1**. To enter matrix [B] for **Expected**, press **MATRX** and then press **2**. Next, press **ENTER** on Calculate, and compare matrix [B] with matrix [A].

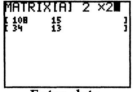
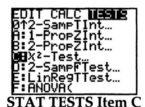

Enter data.　STAT TESTS Item C　Designate matrices.

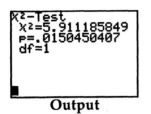
Output

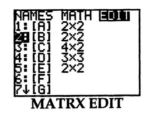
MATRX EDIT

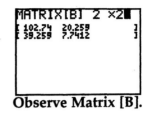
Observe Matrix [B].

If there were no relationship between success and exclusive territory rights, then we would expect 102.74 successful exclusive-territory franchises and 20.259 successful non-exclusive-territory franchises, as shown in the first row of matrix **[B]**. These values differ slightly from the observed values of 108 and 15 in the original data.

In other words, 123 out of 170, or 72.353%, of all franchises are successful. So if there is no relationship between exclusive rights and success, then 72.353% of each type of franchise should be successful. Therefore, the number of successful exclusive-territory franchises would be $(123 / 170) \times 142 = 102.74$.

The χ^2–Test feature also displays the chi-square test statistic, the P-value, and the degrees of freedom for the chi-square test of no association between the row and column variables. In this case, the low P-value of 0.015 gives strong evidence to reject the claim that there is no relationship between exclusive territory rights and franchise success.

Comparison with the 2–PropZTest

The chi-square test for a 2×2 table is equivalent to the two-sided z test for $H_0: p_1 = p_2$ versus $H_a: p_1 \neq p_2$. In Example 9.12, we let p_1 be the true proportion of all successful exclusive-territory franchises and let p_2 be the true proportion of all successful non-exclusive-territory franchises. Then $\hat{p}_1 = 108/142$ and $\hat{p}_2 = 15/28$. If we enter these values into the **2–PropZest** screen and use the alternative $\neq$p2, then we obtain the same P-value of 0.015.

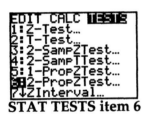
STAT TESTS item 6

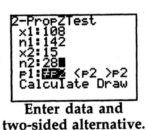
Enter data and two-sided alternative.

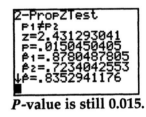
P-value is still 0.015.

9.3 Formulas and Models for Two-Way Tables

In this section, we further explain the test of no association among the row traits and the column traits versus the alternative that there is some relation between these traits.

Example 9.18 The two-way table that follows shows the types of wine purchased in a Northern Ireland supermarket while a certain type of music was being played. Find the conditional distribution of wine given the type of music. Test whether there is an association between the type of wine purchased and the type of music being played.

	Music			
Wine	None	French	Italian	Total
French	30	39	30	99
Italian	11	1	19	31
Other	43	35	35	113
Total	84	75	84	243

Solution. First, enter the data (excluding totals) into a 3×3 matrix **[A]** in the **MATRX EDIT** screen. Next, execute the **TWOWAY** program by entering **3** for both the number of rows and columns. Then observe the conditional distribution of the type of wine given the type of music in matrix **[C]**. We see that the type of music appears to affect the type of wine purchased.

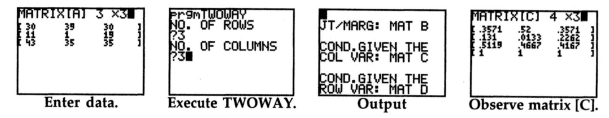

Enter data.   Execute TWOWAY.   Output   Observe matrix [C].

To see if there is an association, we shall test the null hypothesis H_0 that there is no relation between the music and the type of wine purchased. The alternative is that there is a relation, or that the type of wine purchased is dependent upon the music being played.

With the data already entered into matrix **[A]**, we can conduct the test with the χ^2–**Test** feature. We designate matrix **[A]** for **Observed**, and here we shall designate matrix **[E]** for **Expected**. After calculating, we obtain a *P*-value of 0.001 from a test statistic of 18.2792.

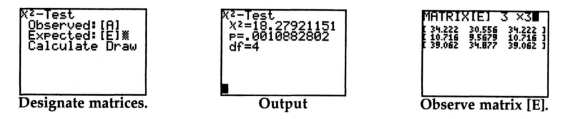

Designate matrices.   Output   Observe matrix [E].

If there were no association between the type of wine purchased and the type of music being played, then there would be only a 0.001 probability of obtaining observed cell counts that differ so much from the expected cell counts shown in matrix **[E]**. Due to this low *P*-value, we can reject the null hypothesis and say that the type of wine purchased is dependent upon the music being played.

Exercise 9.38 A recent study of 865 college students found that 42.5% had student loans. The following table classifies the students by field of study and whether or not they have a loan.

	Student loan	
Field of study	Yes	No
Agriculture	32	35
Child development and family studies	37	50
Engineering	98	137
Liberal arts and education	89	124
Management	24	51
Science	31	29
Technology	57	71

Carry out an analysis to see if there is a relationship between field of study and having a student loan.

Solution. We will test to see if the proportion of students having a student loan is the same regardless of field of study (i.e., if having a loan is independent of field). First, we enter the data into a 7×2 matrix [A] in the **MATRX EDIT** screen. Next, we bring up the χ^2–**Test** screen and designate [A] for **Observed**. Here we shall use matrix [E] for **Expected**. Upon calculating, we obtain a *P*-value of 0.367 from a test statistic of 6.52526.

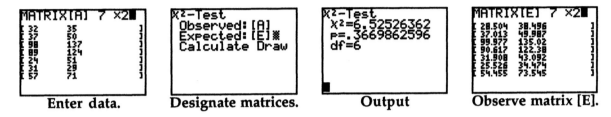

| Enter data. | Designate matrices. | Output | Observe matrix [E]. |

If having a loan were independent of field of study (i.e., if all fields had the same proportion of students with loans), then there would be a 0.367 probability of obtaining observed cell counts that differ so much from the expected cell counts. Because of the high *P*-value, we can say that the observed differences are due to random chance and are not statistically significant. Thus, we will not reject the hypothesis that having a loan is independent of field of study.

For further evidence, we can observe the conditional distribution of students having a loan given the particular fields of study. To do so, we can execute the **TWOWAY** program and view matrix [D]. We see that the percentages having a loan are very close in all fields, except perhaps for those in Management and those in Science.

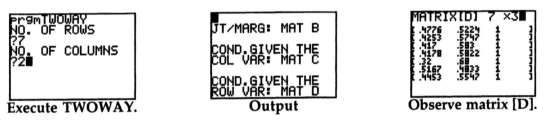

| Execute TWOWAY. | Output | Observe matrix [D]. |

9.4 Goodness of Fit

We conclude with a brief program that performs a goodness of fit test for a specified discrete distribution. Before executing the **FITTEST** program that follows, enter the specified proportions into list **L1** and enter the observed cell counts into list **L2**. The expected cell counts are computed and stored in list **L3**, and the individual contributions to the chi-square test statistic are stored in list **L4**. The program displays the test statistic and the *P*-value.

The FITTEST Program

Program:FITTEST	$:1-\chi^2 cdf(0,X,n-1)\rightarrow P$
:1-Var Stats L$_2$	:ClrHome
:$\Sigma x*L_1 \rightarrow L_3$	:Disp "CHI SQ STAT"
:$(L_2-L_3)^2/L_3 \rightarrow L_4$	:Disp X
:1-Var Stats L$_4$	:Disp "P VALUE"
:$\Sigma x \rightarrow X$	:Disp P

Example 9.24 The following table gives the number of motor vehicle collisions by drivers using a cell phone broken down by days of the week over a 14-month period. Are such accidents equally likely to occur on any day of the week?

Number of collisions by day of the week

Sun.	Mon.	Tue.	Wed.	Thu.	Fri.	Sat.	Total
20	133	126	159	136	113	12	699

Solution. If each day were equally likely, then 1/7 of all accidents should occur on each day. To test the fit of this distribution, we shall use the **FITTEST** program. We first enter 1/7 seven times into list **L1** to specify the expected distribution, and enter the given frequencies from the table into list **L2**. Next, we run the **FITTEST** program to obtain a *P*-value of 0 from a chi-square test statistic of 208.8469.

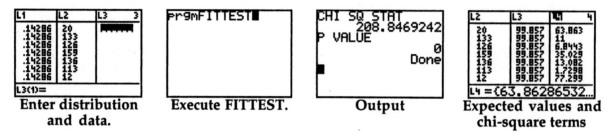

| Enter distribution and data. | Execute FITTEST. | Output | Expected values and chi-square terms |

If these accidents were equally likely to occur on any day of the week, then there would be no chance of obtaining a sample distribution that differs so much from the expected counts of $(1 / 7) \times 699 = 99.857$ for each day. So we can reject the claim that accidents are equally likely on each day. After the program runs, the expected counts are stored in list **L3** and the contributions to the chi-square test statistic from each day are stored in list **L4**.

Exercise 9.47 At a particular college, 29% of undergraduates are in their first year, 27% in their second, 25% in their third, and 19% are in their fourth year. But a random survey found that there were 54, 66, 56, and 30 students in the first, second, third, and fourth years, respectively. Use a goodness of fit test to examine how well the sample reflects the college's population.

Solution. We shall use the **FITTEST** program to test the goodness of fit. Before executing the program, we enter the stated proportions **0.29, 0.27, 0.25,** and **0.19** into list **L1** and the obtained frequencies into list **L2**. Upon running the program, we obtain a *P*-value of 0.17061 from a chi-square test statistic of 5.016. For the given distribution, there is about a 17% chance of obtaining observed sample counts that differ as much as these do from the expected counts (stored in list **L3**).

| Enter distribution and data. | Execute FITTEST. | Output | Expected values and chi-square terms |

CHAPTER

10

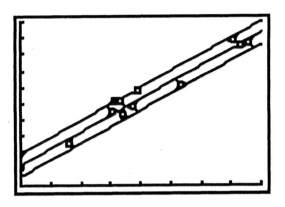

Inference for Regression

10.1 Simple Linear Regression
10.2 More Detail about Simple
Linear Regression

Introduction

In this chapter, we provide details on using the TI-83 Plus to perform the many difficult calculations for linear regression. In particular, we again find and graph the least-squares regression line and compute the correlation. We then can perform a t test to check the hypothesis that the correlation (or, equivalently, the regression slope) is equal to 0. We also provide a program that computes confidence intervals for the regression slope and intercept and another program that computes a prediction interval for a future observation and a confidence interval for a mean response.

10.1 Simple Linear Regression

We begin by demonstrating the **LinRegTTest** feature (item E) from the **STAT TESTS** menu that will compute a least-squares regression line while simultaneously testing the null hypothesis that the regression slope equals 0.

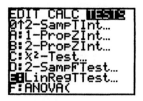

Exercise 10.17 Here are data on the net new money (in billions of dollars) flowing into stock and bond mutual funds from 1985 to 2000.

Year	1985	1986	1987	1988	1989	1990	1991	1992
Stocks	12.8	34.6	28.8	−23.3	8.3	17.1	50.6	97.0
Bonds	100.8	161.8	10.6	−5.8	−1.4	9.2	74.6	87.1

Year	1993	1994	1995	1996	1997	1999	1999	2000
Stocks	151.3	133.6	140.1	238.2	243.5	165.9	194.3	309.0
Bonds	84.6	−72.0	−6.8	3.3	30.0	79.2	−6.2	−48.0

(a) Make a scatterplot with cash flow into stock funds as the explanatory variable. Find the least-squares line for predicting net bond investments from net stock investments.

(b) Is there statistically significant evidence that there is some straight-line relationship between the flows of cash into bond funds and stock funds? State hypotheses, give a test statistic and P-value, and state a conclusion.

Solution. We make a scatterplot as explained in Section 2.1 by entering the data into lists and adjusting the **WINDOW** and **STAT PLOT** settings.

| Enter data into lists. | Adjust WINDOW. | Adjust STAT PLOT. | Press GRAPH. |

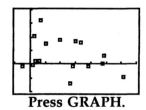

In Chapter 2, we showed how to compute and graph the least-squares line using the **LinReg(a+bx)** command from the **STAT CALC** menu. But now we shall use **LinRegTTest** feature from **STAT TESTS** menu. As before, we first should make sure that the **DiagnosticOn** command has been entered from the **CATALOG**.

With the **LinRegTTest**, we can find the least-squares line while at the same time testing the null hypothesis that the slope β_1 is equal to 0. We will use the alternative hypothesis that $\beta_1 \neq 0$. This test is equivalent to testing the null hypothesis that the correlation ρ is equal to 0 with an alternative that $\rho \neq 0$.

After bringing up the **LinRegTTest** screen, adjust the settings for L1 versus L2 (or to whatever lists contain the data) and enter the alternative ≠0. To enter **Y1** for **RegEQ**, scroll down to the right of **RegEQ**, press **VARS**, scroll right to **Y−VARS**, press **1** for **Function** and then press **1** for **Y1**. Then press **ENTER** on **Calculate**.

We obtain the regression line **y = a + bx** (or $y = \beta_0 + \beta_1 x$), which is also stored into **Y1**. Here the equation rounds to $y = 53.4096 - 0.1962\,x$. Press **GRAPH** to see the plot.

 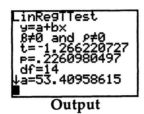  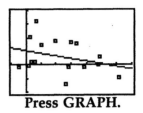

Adjust LinRegTTest.   **Output**   **Scroll down.**   **Press GRAPH.**

The P-value for the t test is given as 0.226 from a t statistic of -1.26622. If β_1 were equal to 0, then there would be a 22.6% chance of obtaining a value for b as low as -0.1962, or of obtaining a correlation as low as $r = -0.32$, with a sample of this size. This rather high P-value means that we do *not* have statistically significant evidence that there is some straight-line relationship between the flows of cash into bond funds and stock funds. In other words, we do not have enough evidence to reject that $\beta_1 = 0$ or to reject that $\rho = 0$.

Confidence Intervals for Slope and Intercept

We now provide a program that will compute confidence intervals for the slope and for the intercept of the linear regression model $y = \beta_0 + \beta_1 x$. Before executing the **REG1** program below, we must enter paired data must into lists **L1** and **L2**.

The REG1 Program

```
PROGRAM:REG1
:Disp "CONF. LEVEL"        :ClrHome
:Input R                   :Disp "INTCPT.INTERVAL"
:LinRegTTest L₁,L₂,1       :Disp a-TA
:"tcdf(0,X,n-2)"→Y₂        :Disp a+TA
:solve(Y₂-R/2,X,2)→T       :Disp "SLOPE INTERVAL"
:s/√(nσx²)→B               :Disp b-TB
:s√(1/n+x̄²/(nσx²))→A       :Disp b+TB
```

Example Using the data from Exercise 10.17, find 90% confidence intervals for the slope β_1 and the intercept β_0 of the linear regression model.

Solution. After entering the data into lists **L1** and **L2**, bring up the **REG1** program and enter .90 for **CONF. LEVEL**. The interval {12.9127, 93.9065} for the intercept and the interval {−0.4692, 0.0767} for the slope are displayed.

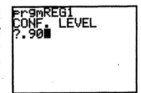 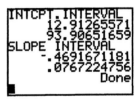

Mean Response and Prediction Confidence Intervals

We now provide another program, **REG2**, which computes a confidence interval for a mean response or a prediction interval for an estimated response. Before this program can be executed, paired data must be entered into lists **L1** and **L2**.

The REG2 Program

```
PROGRAM:REG2                          :s√(1/n+(X-X̄)²/(nσx²))→E
:Disp "1 = MEAN INT.","2 = PRED. INT." :s√(1+1/n+(W-X̄)²/(nσx²))→D
:Input C                              :a+bX→Y
:If C=1                               :a+bW→Z
:Then                                 :ClrHome
:Disp "MEAN VALUE OF X?"              :If C=1
:Input X                             :Then
:Else                                :Disp "MEAN INTERVAL"
:Disp "FUT. VAL. OF X?"              :Disp Y-TE
:Input W                             :Disp Y+TE
:End                                 :Else
:Disp "CONF. LEVEL"                  :Disp "PREDICTION INT."
:Input R                             :Disp Z-TD
:LinRegTTest L₁,L₂,1                  :Disp Z+TD
:"tcdf(0,X,n-2)"→Y₂                   :End
:solve(Y₂-R/2,X,2)→T
```

Example With the data from Exercise 10.17, find a 90% confidence interval for the mean net bond investment for a year with an average of $120 billion net stock investment. Find a 95% prediction interval for a future year with an average of $170 billion net bond investment.

Solution. If the data have been entered into lists **L1** and **L2**, then we can compute the intervals separately by entering either **1** or **2** when prompted in the **REG2** program. For the first interval, first enter **1**, then **120**, then the confidence level of **.9**. For the prediction interval, first enter **2**, then **170**, then the confidence level of **.95**.

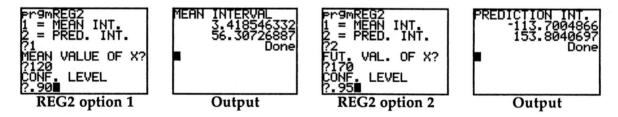

| REG2 option 1 | Output | REG2 option 2 | Output |

Exercise 10.23 The table below gives the amount of lean in tenths of a millimeter in excess of 2.9 meters for the Leaning Tower of Pisa from 1975 to 1987.

Year	75	76	77	78	79	80	81	82	83	84	85	86	87
Lean	642	644	656	667	673	688	696	698	713	717	725	742	757

(a) Plot the data.
(b) What is the equation of the least-squares line? What percentage of the variation in lean is explained by this line?
(c) Give a 99% confidence interval for the average rate of change of the lean.

Exercise 10.24 Using the least-squares equation from Exercise 10.23, calculate a predicted value for the lean in 1918.

Exercise 10.25 Use the least-squares equation from Exercise 10.23 to predict the tower's lean in the year 2007. Give the margin of error for a 99% prediction interval for 2007.

Solutions. 10.23 (a) Enter the data into lists **L1** and **L2**, set an appropriate window, adjust the **STAT PLOT** settings for a scatterplot, and graph. The trend in lean appears to be linear in time with a positive slope β_1.

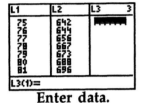  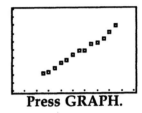

| Enter data. | Adjust WINDOW. | Adjust STAT PLOT. | Press GRAPH. |

(b) We now can find the equation of the least-squares line while at the same time testing the null hypothesis H_0: $\beta_1 = 0$ (or $\rho = 0$), with an alternative H_a: $\beta_1 > 0$ (or $\rho > 0$). We simply bring up the **LinRegTTest** screen, adjust the list settings and alternative, and calculate.

 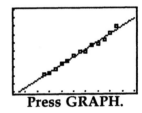

| Adjust LinRegTTest. | Output | Scroll down. | Press GRAPH. |

We obtain a least-squares line of $y = -61.12 + 9.3187\,x$ with $r^2 \approx 0.988$. Thus, about 98.8% of the variation in lean is explained by this least-squares line. For the hypothesis test, we obtain a P-value of $3.25 \times 10^{-12} \approx 0$. This low value gives strong evidence to reject H_0 and conclude that $\beta_1 > 0$ (or $\rho > 0$), which means that there is a positive correlation.

(c) With the data entered into into lists **L1** and **L2**, we can use the **REG1** program to find a confidence interval for the average rate of change of the lean (i.e., for the slope of the least-squares line). We simply enter .99 for **CONF. LEVEL** in the program, and obtain the interval for the slope as (8.3561, 10.2812).

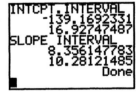

10.24: Because 18 is the coded value for the year 1918, we simply evaluate the least-squares line at 18. The **LinRegTTest** feature has stored the line into function **Y1**; so we simply retrieve **Y1** from the **VARS Y–VARS FUNCTION** screen and evaluate **Y1(18)** as 106.615.

 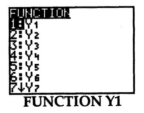 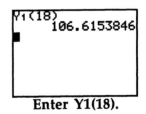

| VARS | Y–VARS | FUNCTION Y1 | Enter Y1(18). |

10:25: We now use 107 as the coded value for the year 2007. We first evaluate the predicted lean with **Y1(107)**, and then find a 99% prediction interval with option 2 of the **REG2** program.

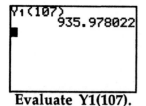

Evaluate Y1(107).

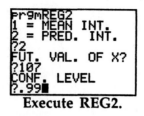

Execute REG2.

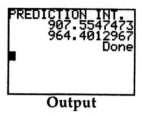

Output

The prediction interval is given as [907.5547, 964.4013]. Because the midpoint of the prediction interval equals the value of the regression line evaluated at the specified future value of $x = 107$, the margin of error is found by $935.9780 - 907.5547 = 28.4233$.

10.2   More Detail about Simple Linear Regression

Analysis of variance (ANOVA) is another method to test the null hypothesis $H_0: \beta_1 = 0$, with an alternative $H_a: \beta_1 \neq 0$. The **REG3** program that follows performs such a test, stores an ANOVA table into lists **L4, L5,** and **L6,** and displays the associated F-statistic and P-value. Before executing the **REG3** program, we must enter paired data into lists **L1** and **L2**.

Note: The ANOVA table generated by the **REG3** program is not the same one that will be generated by the TI-83's built-in **ANOVA(** command that is used for testing whether or not several populations have identical means. That command will be explained in Chapter 12.

The REG3 Program for Linear Regression ANOVA

```
PROGRAM:REG3                          :1→L₄(1):A→L₅(1):A→L₆(1)
:LinReg(a+bx) L₁,L₂                   :n-2→L₄(2):B→L₅(2):B/(n-2)→L₆(2)
:sum(seq((a+bL₁(I)-ȳ)²,              :n-1→L₄(3):C→L₅(3):C/(n-1)→L₆(3)
 I,1,dim(L₁)))→A                      :ClrHome
:sum(seq((L RESID(I))²,I,1,           :Disp "F-STAT, P-VAL"
 dim(L₁)))→B                          :Output(2,2,{round(F,3),round(P,4)})
:sum(seq((L₂(I)-ȳ)²,                  :Output(4,1,"SEE L4, L5, L6")
 I,1,dim(L₁)))→C                      :Output(5,5,"DF, SS, MS")
:(n-2)A/B→F                           :Output(6,3,"M")
:1-Fcdf(0,F,1,n-2)→P                  :Output(7,3,"E")
:ClrList L₄,L₅,L₆                     :Output(8,3,"T")
```

Example Consider the data from Exercise 10.17 on the net new money (in billions of dollars) flowing into stock and bond mutual funds from 1985 to 2000.

Year	1985	1986	1987	1988	1989	1990	1991	1992
Stocks	12.8	34.6	28.8	−23.3	8.3	17.1	50.6	97.0
Bonds	100.8	161.8	10.6	−5.8	−1.4	9.2	74.6	87.1

Year	1993	1994	1995	1996	1997	1999	1999	2000
Stocks	151.3	133.6	140.1	238.2	243.5	165.9	194.3	309.0
Bonds	84.6	−72.0	−6.8	3.3	30.0	79.2	−6.2	−48.0

(a) Construct the ANOVA table.
(b) State and test the hypotheses using the ANOVA F-statistic.
(c) Give the degrees of freedom for the F-statistic for the test of H_0.
(d) Verify that the square of the t statistic for the equivalent t test is equal to the F-statistic in the ANOVA table.

Solution. (a) To construct an ANOVA table with the **REG3** program, we first must enter our data into lists **L1** (for stocks) and **L2** (for bonds). After doing so, bring up and execute the program. (There is nothing to input, so just press **ENTER**.)

The program displays the F-statistic and P-value. The ANOVA table is stored in lists **L4**, **L5**, and **L6** and contains the degrees of freedom (DF), the sum of squares (SS), and the mean square (MS) for each of the model (M), error (E), and total (T).

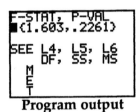
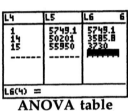

Program output   ANOVA table

(b) The ANOVA test is about the linear regression slope β_1. We test the null hypothesis H_0: $\beta_1 = 0$ with the alternative H_a: $\beta_1 \neq 0$. With the above P-value of 0.2261, we do not have strong evidence in this case to reject H_0 in favor of the alternative.

(c) The degrees of freedom for the F-statistic is given by $n - 2$, which in this case is 14. This number is the same as the degrees of freedom of the error (E) displayed in the ANOVA table.

(d) The **LinRegTTest** was applied in Exercise 10.17 and the results are displayed below. The t statistic for the t test was computed as $t = -1.266220727$. If we square this value, then we obtain 1.603314929, which is the actual value of the displayed rounded-off F-statistic from the ANOVA test.

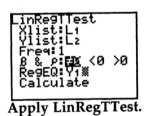
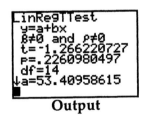

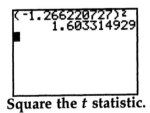

Apply LinRegTTest.   Output   Scroll down.   Square the t statistic.

Sample Correlation and the *t* Test

One may be required to perform a correlation *t* test without an actual data set, but instead by using only the values of the sample correlation *r* and the sample size *n*. Although the TI-83 Plus does not have a built-in procedure for this type of test, the *t* statistic and *P*-value are easily calculated in this case. Here the test statistic, which follows a $t(n-2)$ distribution, is given by

$$t = \frac{r\sqrt{n-2}}{\sqrt{1-r^2}}$$

Exercise 10.42 In a study of 564 children who were 2 to 6 years of age, the relationship of food neophobia and the frequency of consumption gave a correlation of $r = -0.15$ for meat. Perform a significance test about the correlation of meat neophobia and the frequency of meat consumption among children 2 to 6 years of age.

Solution. We shall test $H_0: \rho = 0$ versus $H_a: \rho < 0$. The rejection region is a left tail; thus the *P*-value will be the left-tail probability of the $t(n-2) = t(562)$ distribution. We compute the *t* statistic and *P*-value "manually" by entering $-.15*\sqrt{(562)}/\sqrt{(1-.15^2)}$ to obtain $t \approx -3.59667$. Next, we compute the *P*-value $P(t(562) \le -3.59667)$ with the **tcdf(** command from the **DISTR** menu. Here we enter **tcdf(–1E99, –3.59667, 562)** to obtain a *P*-value of 1.75×10^{-4}.

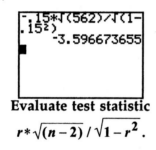

Evaluate test statistic   **Compute P-value.**
$r*\sqrt{(n-2)}/\sqrt{1-r^2}$.

   If the true correlation were 0, then there would be only a 0.000175 probability of obtaining a sample correlation as low as $r = -0.15$ with a random sample of size 564. We have statistical evidence to reject H_0 in favor of the alternative that $\rho < 0$.

   We note that for a two-sided alternative, then our *P*-value would be $2 \times (1.75 \times 10^{-4}) = 3.5 \times 10^{-4}$, which would still be small enough for us to reject H_0.

CHAPTER
11

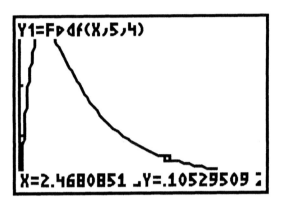

Multiple Regression

11.1 Inference for Multiple Regression
11.2 A Case Study

Introduction

In this chapter, we demonstrate how to use a program for the TI-83 Plus to calculate a multiple linear regression model.

11.1   Inference for Multiple Regression

The **MULTREG** program below computes the regression coefficients and an ANOVA table for the multiple linear regression model $\mu_y = \beta_0 + \beta_1 x_1 + \ldots + \beta_p x_p$. The squared correlation coefficient, the F-statistic, the P-value, and the standard deviation are also displayed.

The MULTREG Program

```
PROGRAM:MULTREG
:Disp "NO. OF VAR."
:Input M
:Disp "NO. OF SAMPLES"
:Input N
:{M,M}→dim([B])
:{M,1}→dim([C])
:sum(seq([A](I,M),I,1,N))→
 [C](1,1)
:For(J,1,M-1)
:sum(seq([A](I,J)[A](I,M),I,1,N))
 →[C](J+1,1)
:End
:N→[B](1,1)
:For(J,1,M-1)
:sum(seq([A](I,J),I,1,N))→
 [B](1,J+1)
:End
:For(I,1,M-1)
:[B](1,I+1)→[B](I+1,1)
:End
:For(J,1,M-1)
:For(K,1,M-1)
:sum(seq([A](I,J)[A](I,K),I,1,N))
 →[B](J+1,K+1)
:End
:End
:[B]⁻¹*[C]→[D]
:[C](1,1)/N→Y
:{N,1}→dim([E])
```

```
:For(I,1,N)
:[D](1,1)+sum(seq([D](J+1,1)[A](I,J),
 J,1,M-1))→[E](I,1)
:End
:sum(seq(([E](I,1)-Y)²,I,1,N))→A
:sum(seq(([A](I,M)- [E](I,1))²,I,1,N))→B
:sum(seq(([A](I,M)-Y)²,I,1,N))→C
:(A/(M-1))/(B/(N-M))→F
:1-Fcdf(0,F,M-1,N-M)→P
:{M-1,N-M,N-1}→L₁
:{A,B,C}→L₂
:L₂/L₁→L₃
:sum(seq(([A](I,M)-[E](I,1))²,I,1,N))→R
:ClrHome
:Disp "DF,SS,MS (M,E,T)"
:Disp {M-1,round(A,3),round(A/
 (M-1),3)}
:Disp {N-M,round(B,3),round(B/
 (N-M),3)}
:Disp {N-1,round(C,3),round(C/
 (N-1),3)}
:Output(5,2,"R²")
:Output(5,5,round(A/C,9))
:Output(6,2,"F")
:Output(6,5,round(F,9))
:Output(7,2,"P")
:Output(7,5,round(P,9))
:Output(8,2,"S")
:Output(8,5,round(√(R/(N-M)),9))
```

To execute the program, we first must enter our sample data into matrix **[A]**, and we must specify how many variables are being used, which includes the several independent variables and the dependent variable. For example, the model $\mu_y = \beta_0 + \beta_1 x_1 + \beta_2 x_2 + \beta_3 x_3 + \beta_4 x_4 + \beta_5 x_5$ has six variables, one of which is being predicted from the other five. The displayed ANOVA table is also stored into lists **L1**, **L2**, and **L3**. The program stores the regression coefficients of the model in matrix **[D]**.

11.2   A Case Study

Example Consider the CSDATA of a sample of 24 students at a large university that uses a 4.0 GPA grade scale. Run a multiple regression analysis for predicting the GPA from the three high school grade variables.

OBS	GPA	HSM	HSS	HSE	SATM
1	3.32	10	10	10	670
2	2.26	6	8	5	700
3	2.35	8	6	8	640
4	2.08	9	10	7	670
5	3.38	8	9	8	540
6	3.29	10	8	8	760
7	3.21	8	8	7	600
8	2.00	3	7	6	460
9	3.18	9	10	8	670
10	2.34	7	7	6	570
11	3.08	9	10	6	491
12	3.34	5	9	7	600
13	1.40	6	8	8	510
14	1.43	10	9	9	750
15	2.48	8	9	6	650
16	3.73	10	10	9	720
17	3.80	10	10	9	760
18	4.00	9	9	8	800
19	2.00	9	6	5	640
20	3.74	9	10	9	750
21	2.32	9	7	8	520
22	2.79	8	8	7	610
23	3.21	7	9	8	505
24	3.08	9	10	8	559

Solution. We use the model $\mu_{GPA} = \beta_0 + \beta_1 \times HSM + \beta_2 \times HSS + \beta_3 \times HSE$, which is of the form $\mu_y = \beta_0 + \beta_1 x_1 + \beta_2 x_2 + \beta_3 x_3$. In this case, there are 24 sample points and four variables.

We must first enter the data into matrix **[A]**, much as we have been entering data into lists. Matrix **[A]** should have 24 rows and 4 columns. However, to execute the program correctly, the dependent variable must be in the *last* column.

In the **MATRX EDIT** screen, enter the dimensions of matrix **[A]** as 24 × 4, then enter the data. The TI-83 Plus enters data into a matrix by going sequentially across the rows. So the first row will be   10   10   10   3.32, and the second row will be 6   8   5   2.26. After the data is entered into matrix **[A]**, execute the **MULTREG** program.

MATRX EDIT

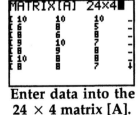

Enter data into the 24 × 4 matrix [A].

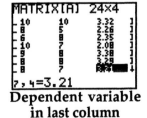

Dependent variable in last column

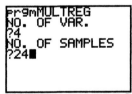

Execute MULTREG.

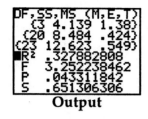
Output

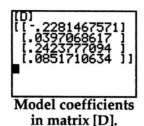
Model coefficients
in matrix [D].

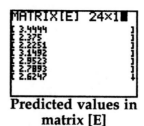
Predicted values in
matrix [E]

Scroll down.

The displayed ANOVA table gives the degrees of freedom (DF), the sum of squares (SS), and the mean square (MS) for each of the model (M), error (E), and total (T). The table is also stored into lists **L1**, **L2**, and **L3** in the **STAT EDIT** screen. The squared correlation coefficient, the F-statistic and P-value of the F-test, and the standard deviation S are also displayed. The value of R^2 here is telling us that only 32.8% of the GPAs are explained by the high school variables.

The F-test is for the null hypothesis that each variable coefficient of the linear regression model is equal to 0: H_0: $\beta_1 = \beta_2 = \beta_3 = 0$. In this case, the P-value is 0.0433, which is a value usually considered low enough to be statistically significant. Thus, even with this small sample of size 24, we have enough evidence to reject H_0.

The program stores the computed regression coefficients in matrix [D]. In this case, our model becomes $\mu_{GPA} \approx -0.228 + 0.0397 \times \text{HSM} + 0.242377 \times \text{HSS} + 0.085171 \times \text{HSE}$. Finally, the program stores the model's predicted values for each sample point in matrix [E].

Example   Use the preceding regression model to predict the GPA of a student with HSM = 8, HSS = 7, and HSE = 10.

Solution. We must evaluate $-0.228 + 0.0397 \times \text{HSM} + 0.242377 \times \text{HSS} + 0.085171 \times \text{HSE}$ for these variables, which is easy enough to compute on the **Home** screen. However, if we want a more accurate (non-rounded) value, then we can perform matrix multiplication with the regression coefficients stored in matrix [D].

To multiply the matrices, we will enter the given variables into matrix [F]. But we must include an additional 1 to account for the slope. So we will enter the variables {1, 8, 7, 10} into a 1×4 matrix [F] in the **MATRX EDIT** screen. After entering the values into [F], return to the **Home** screen and multiply [F]*[D]. (Retrieve the matrices from the **MATRX NAMES** screen). The predicted GPA is about 2.64.

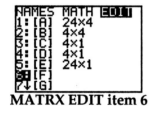
MATRX EDIT item 6

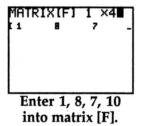
Enter 1, 8, 7, 10
into matrix [F].

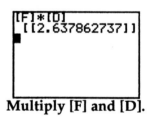
Multiply [F] and [D].

Example   With the above 24 sample points from the CSDATA, perform regression analysis for predicting the SATM from the three high school grade variables. Then find the predicted SATM for a student with HSM = 9, HSS = 6, and HSE = 8.

Solution. First, we must edit the last column of matrix [A] so that it contains the SATM scores. Then we can rerun the **MULTREG** program to obtain a new regression model in matrix [D].

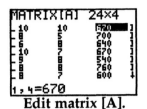
Edit matrix [A].

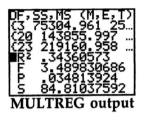
MULTREG output

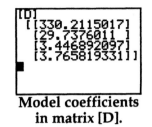
Model coefficients
in matrix [D].

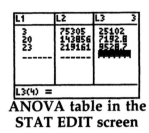
ANOVA table in the
STAT EDIT screen

The new model is μ_{SATM} = 330.2115 + 29.7376 × HSM + 3.4469 × HSS + 3.7658 × HSE. We note that the displays of the mean square errors in the ANOVA table have been truncated from the **Home** screen; however, they are easily recovered from the **STAT EDIT** screen.

The displayed P-value of 0.0348 gives us statistical evidence to reject that all regression coefficients of HSM, HSS, and HSE are 0. Thus, at least one coefficient is non-zero, and its parameter is correlated to SATM scores. If we store {1, 9, 6, 8} in the 1×4 matrix [F] and multiply [F]*[D], then we obtain a predicted SATM score of about 649.

Exercise 11.20 Below are some data for the top ten Internet brokerages.

ID	Broker	Mshare	Accts	Assets
1	Charles Schwab	27.5	2500	219.0
2	E*Trade	12.9	909	21.1
3	TD Waterhouse	11.6	615	38.8
4	Datek	10.0	205	5.5
5	Fidelity	9.3	2300	160.0
6	Ameritrade	8.4	428	19.5
7	DLJ Direct	3.6	590	11.2
8	Discover	2.8	134	5.9
9	Suretrade	2.2	130	1.3
10	National Discount Brokers	1.3	125	6.8

(a) Use a simple linear regression to predict assets using the number of accounts. Give the regression equation and the results of the significance test for the regression coefficient.
(b) Do the same using market share to predict assets.
(c) Run a multiple regression using both the number of accounts and market share to predict assets. Give the multiple regression equation and the results of the significance test for the two regression coefficients.
(d) Compare the results of parts (a), (b), and (c).

Solution. (a) We shall use the **LinRegTTest** feature (item E in **STAT TESTS** menu). To do so, we will enter the data into lists **L4**, **L5**, and **L6**. Then we execute the **LinRegTTest** on lists **L5** and **L6** to obtain a linear regression equation of $y = a + bx \approx -17.1215 + 0.0832x$ for predicting assets using the number of accounts.

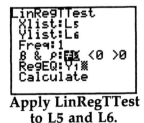
Enter data into lists
L4, L5, and L6.

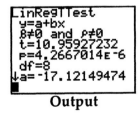
Apply LinRegTTest
to L5 and L6.

Output

Scroll down.

The P-value for a two-sided significance test is 4.2667×10^{-6} If the regression coefficient β_1 were equal to 0, then there would be only a very small chance of obtaining a value of b as large as 0.0832, or a correlation as high as 0.96827, even with such a small sample. Thus, we have strong evidence to reject the hypothesis that $\beta = 0$.

(b) Now we execute the **LinRegTTest** on lists **L4** and **L6** to obtain a linear regression equation of $y = a + bx \approx -19.9 + 7.68x$ for predicting assets using market share.

 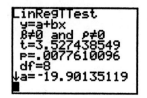 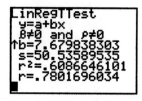

The P-value for the two-sided significance test is 0.007761. If the regression coefficient β_1 were equal to 0, then there would be less than a 1% chance of obtaining a value of b as high as 7.68 or a correlation as high as 0.78. Thus, we again can reject the hypothesis that $\beta = 0$.

(c) To execute the **MULTREG** program, we must first move the data to matrix **[A]**. To do so, we can use the **List▶matr(** command from the **MATRX MATH** menu. After bringing this command to the **Home** screen, enter the command **List▶matr(L4,L5,L6,[A])**, where **[A]** is retrieved from the **MATRX NAMES** menu.

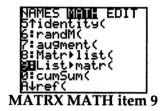 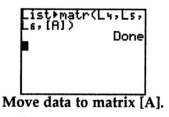 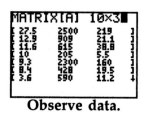
MATRX MATH item 9   **Move data to matrix [A].**   **Observe data.**

Lastly, execute the **MULTREG** program and bring up matrix **[D]** to see the regression coefficients:

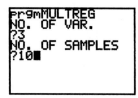 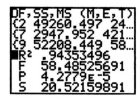 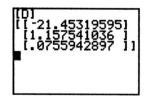

We obtain a multiple linear regression equation of

$$\mu_{Assets} = -21.4532 + 1.15754 \times Mshare + 0.0756 \times Accts$$

The P-value of (approximately) 0 gives significant evidence to reject the null hypothesis that $\beta_1 = \beta_2 = 0$.

(d) Comparing the r^2 values from each part, we see that the multiple linear regression provides the best "fit" with 94.35% of the assets being explained by the other two variables.

CHAPTER
12

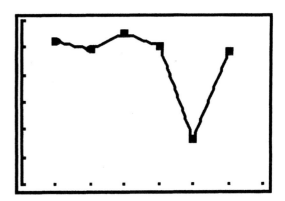

One-Way Analysis of Variance

Introduction

In this chapter, we perform one-way analysis of variance (ANOVA) to test whether several normal populations, assumed to have the same variance, also have the same mean.

12.1 Inference for One-Way Analysis of Variance

We begin with an exercise that demonstrates the TI-83's built-in analysis of variance **ANOVA(** command from the **STAT TESTS** menu (item F).

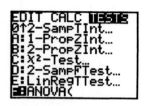

Exercise 12.27 The data below give the lengths in millimeters of three varieties of the tropical flower *Heliconia*, which are fertilized by different species of hummingbird on the island of Dominica. Perform an ANOVA test to compare the mean lengths of the flowers for the three species.

H. bihai

47.12	46.75	46.81	47.12	46.67	47.43	46.44	46.64
48.07	48.34	48.15	50.26	50.12	46.34	46.94	48.36

H. caribaea red

41.90	42.01	41.93	43.09	41.47	41.69	39.78	40.57
39.63	42.18	40.66	37.87	39.16	37.40	38.20	38.07
38.10	37.97	38.79	38.23	38.87	37.78	38.01	

H. caribaea yellow

36.78	37.02	36.52	36.11	36.03	35.45	38.13	37.1
35.17	36.82	36.66	35.68	36.03	34.57	34.63	

Solution. The built-in **ANOVA(** command requires the data to be in lists. Here we shall enter the data into lists **L1**, **L2**, and **L3**. After doing so, we evaluate the basic statistics of each list.

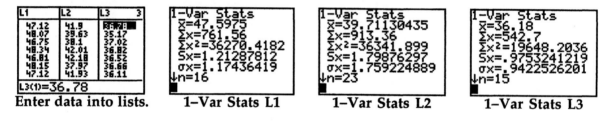

Enter data into lists.   1–Var Stats L1   1–Var Stats L2   1–Var Stats L3

Next, we test the hypothesis that the mean lengths of the three species are equal: H_0: $\mu_1 = \mu_2 = \mu_3$. To do so, we bring the **ANOVA(** command to the **Home** screen and enter the command **ANOVA(L1, L2, L3)**.

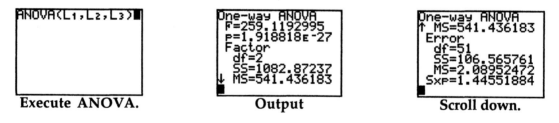

Execute ANOVA.   Output   Scroll down.

We receive a P-value of 1.92×10^{-27} from an F-statistic of 259.1193. The pooled deviation value is also displayed as **Sxp** ≈ 1.4455. If the true means were equal, then there would be almost no chance of the sample means varying by as much as they do with samples of these sizes. Thus, we have significant evidence to reject the claim that the mean lengths of these species are equal. We note that the R^2 value is not displayed, but it can be computed from the two displayed SS values. Here we can use $R^2 =$ SSF/(SSF + SSE) $= 1082.87237/(1082.87237 + 106.565761) \approx 0.9104$.

Because we have rejected the hypothesis that the means lengths are all equal, we can say that there is at least one pair of species that have different means. From the summary statistics, it appears that the species *H. bihai* and *H. caribaea yellow* have different mean lengths. But the sample means of *H. caribaea red* and *H. caribaea yellow* are close enough so that one might hypothesize that these species have the same mean length.

We can test any pair of species for equality of mean very quickly using the **2–SampTTest** screen. We demonstrate below with the pairs (**L1, L2**) and (**L2, L3**). Due to the extremely low P-values in the outputs, we see that we can reject both that $\mu_1 = \mu_3$ and that $\mu_2 = \mu_3$.

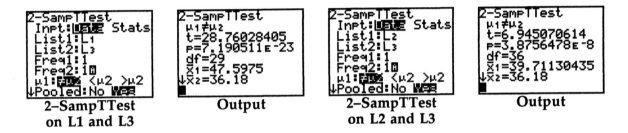

| 2–SampTTest on L1 and L3 | Output | 2–SampTTest on L2 and L3 | Output |

ANOVA for Summary Statistics

When raw data are given, then we can enter the data into lists and use the built-in **ANOVA(** command to test for equality of means. However, sometimes the summary statistics are given instead. In this case, we can use the **ANOVA1** program below to perform the analysis of variance.

The ANOVA1 Program

```
PROGRAM:ANOVA1                                    :Output(3,1,"SP")
:1-Var Stats L₂,L₁                                :Output(3,6,S)
:sum(seq(L₁(I)(L₂(I)-x̄)²,I,1,dim(L₁)))→A          :Output(4,1,"MSG")
:sum(seq((L₁(I)-1)(L₃(I)²),I,1,dim(L₁)))→B         :Output(4,6,A/(I-1))
:dim(L₁)→I                                         :Output(5,1,"MSE")
:(A/(I-1))/(B/(n-I))→F                             :Output(5,6,B/(n-I))
:1-Fcdf(0,F,I-1,n-I)→P                             :Output(6,1,"R^2")
:√(sum(seq((L₁(J)-1)*L₃(J)²,J,1,dim(L₁)))          :Output(6,6,A/(A+B))
 /(n-I))→S                                         :Output(7,1,"F")
:ClrHome                                           :Output(7,6,F)
:Output(2,1,"MEAN")                                :Output(8,1,"P")
:Output(2,6,x̄)                                     :Output(8,6,round(P,8))
```

The program computes and displays the overall sample mean, the pooled sample deviation s_p, the mean square for groups MSG, the mean square for error MSE, the coefficient of determination, the F-statistic, and the P-value. Before executing the program, enter the sample sizes into list **L1**, the sample means into list **L2**, and the sample deviations into list **L3**.

Example 12.3 The table below gives the summary data on safety climate index (SCI) as rated by workers during a study on workplace safety. Apply the ANOVA test and give the pooled deviation, the value of R^2, the F-statistic, and the P-value.

Job category	n	$\bar{x}$	s
Unskilled workers	448	70.42	18.27
Skilled workers	91	71.21	18.83
Supervisors	51	80.51	14.58

Solution. Enter the summary statistics into lists **L1**, **L2**, and **L3**, and execute the **ANOVA1** program. The desired statistics are then displayed. In particular, we obtain a pooled deviation of $s_p = 18.07355$ and a coefficient of determination of $R^2 \approx 0.023756$.

Enter values of n, $\bar{x}$, and s into lists **L1**, **L2**, and **L3**.   Execute ANOVA.

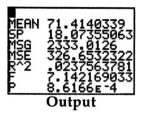

Output

Due to the low P-value of 0.00086, we see that there is strong evidence to reject the claim that each group of workers has the same mean SCI.

12.2   Comparing the Means

In this section, we provide a program for analyzing population contrasts based on summary statistics. Before executing the **CONTRAST** program, enter the sample sizes into list **L1**, the sample means into list **L2**, the sample deviations into list **L3**, and the contrast equation coefficients into list **L4**. When prompted during the program, enter either **1** or **2** for a one-sided or two-sided alternative. The program then displays the P-value for a significance test and a confidence interval for mean contrast.

The CONTRAST Program

PROGRAM:CONTRAST	:0.5-tcdf(0,abs(T),D)→P
:Disp "ALTERNATIVE"	:ClrHome
:Input A	:Disp "P-VALUE"
:Disp "CONF. LEVEL"	:If A=1
:Input R	:Then
:1-Var Stats L₂,L₁	:Disp P
:√(sum(seq((L₁(I)-1)*L₃(I)²,	:Else
I,1,dim(L₁)))/(n-dim(L₁)))→S	:Disp 2*P
:sum(seq(L₂(I)L₄(I),I,1,dim(L₁)))→C	:End
:S√(sum(seq(L₄(I)²/L₁(I),I,1,	:"tcdf(0,X,D)"→Y₁
dim(L₁))))→E	:solve(Y₁-R/2,X,2)→Q
:C/E→T	:Disp "CONF.INTERVAL"
:1-Var Stats L₁	:Disp {round(C-QE,3),
:Σx-dim(L₁)→D	round(C+QE,3)}

Example 12.15 Using the data from Example 12.3, perform a two-sided significance test for the contrast between the average SCI of supervisors compared with the other two groups of workers. Also, find a 95% confidence interval for the contrast.

Solution. We use the contrast equation $\Psi = -\frac{1}{2}\mu_{UN} - \frac{1}{2}\mu_{SK} + 1\mu_{SU}$, and test the hypothesis $H_0: \mu_{SU} = \frac{1}{2}\mu_{UN} + \frac{1}{2}\mu_{SK}$ with the alternative $H_a: \mu_{SU} \neq \frac{1}{2}\mu_{UN} + \frac{1}{2}\mu_{SK}$.

If we still have the sample sizes, sample means, and sample deviations entered into lists **L1**, **L2**, and **L3**, then we just need to add the coefficients of Ψ into list **L4**. Then we execute the **CONTRAST** program by entering 2 for two-sided alternative and entering the desired confidence level.

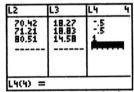

Enter summary statistics into lists L1, L2, and L3, and coefficients of Ψ into L4.

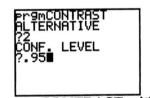

Execute CONTRAST with a two-sided alternative, and a 95% confidence level.

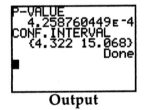

Output

With the low *P*-value of 0.000425876, we see that there is a significant contrast. According to the displayed confidence interval, the mean SCI rating by supervisors could be from 4.322 higher to 15.068 higher than the average of the mean SCI ratings by unskilled workers and skilled workers.

The Power of the ANOVA Test

Lastly, we provide a supplementary program to compute the *approximate* power of the ANOVA test under the alternative H_a that the true population means are $\mu_1, \mu_2, \ldots, \mu_I$. The power is computed by using a Winer approximation of the noncentral F distribution.

The ANPOWER Program

```
PROGRAM:ANPOWER                        :(√((2U-1)*(U/V)*C)-√(2(U+L)-
:Disp "LEVEL OF SIG?"                   (U+2L)/(U+L)))/√((U+2L)/(U+L)+
:Input A                                (U/V)C)→Z
:Disp "STANDARD DEV?"                  :normalcdf(Z,1E99,0,1)→P
:Input S                               :ClrHome
:sum(seq(L₁(I),I,1,dim(L₁)))→N         :Output(1,2,"F*")
:sum(seq(L₁(I)*L₂(I),I,1,dim(L₁)))/    :Output(1,5,C)
 N→M                                   :Output(2,1,"DFG")
:sum(seq(L₁(I)*(L₂(I)-M)^2,            :Output(2,5,U)
 I,1,dim(L₁)))/S^2→L                   :Output(3,1,"DFE")
:dim(L₁)-1→U                           :Output(3,5,V)
:N-dim(L₁)→V                           :Output(4,1,"L")
:"Fcdf(0,X,U,V)"→Y₁                    :Output(4,5,L)
:solve(Y₁-(1-A),X,5(UV-2V)/            :Output(6,1,"PWR")
 (UV+2V))→C                            :Output(6,5,P)
```

Before executing the **ANPOWER** program, enter the successive sample sizes $n_1, \ldots, n_I$ into list **L1** and the alternative population means $\mu_1, \ldots, \mu_I$ into list **L2**. Then enter the level of significance and the guessed standard deviation when prompted while running the program. The approximate power is displayed along with the values of F^*, DFG, DFE, and the noncentrality parameter λ.

Example 12.27 A reading comprehension study had 10 subjects in each of three groups. Estimate the power for the alternative $\mu_1 = 41$, $\mu_2 = 47$, $\mu_3 = 44$, with $\sigma = 7$, at the 5% level of significance.

Solution. First, we enter the three sample sizes of 10 into list **L1** and the alternative means into list **L2**. Then we execute the **ANPOWER** program to obtain an approximate power of 0.35436.

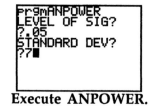

Enter sample sizes into L1 and
alternative means into L2.

Execute ANPOWER.

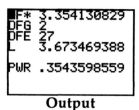

Output

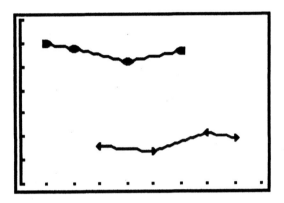

Two-Way
Analysis of
Variance

Introduction

In this chapter, we provide a program that performs two-way analysis of variance to test for equality of means simultaneously among populations and traits in a two-factor experiment.

13.1   The Two-Way ANOVA Model

Example 13.9 The table below gives the mean length of time (in minutes) that various groups of people spent eating lunch in various settings. Plot the group means for this example.

Number of people eating

Lunch setting	1	2	3	4	≥5
Workplace	12.6	23.0	33.0	41.1	44.0
Fast food restaurant	10.7	18.2	18.4	19.7	21.9

Solution. We note that we cannot use ANOVA to analyze this data because each cell contains the *mean* of an unknown number of subjects, and we cannot assume that there is a fixed number of measurements per cell. In order to perform two-way ANOVA, we also would need to know the sample sizes and the variances of the data that produced the mean of each cell. However, we can plot the means with a time plot in order to observe a possible interaction.

First, enter the integers 1 through 5 into list **L1**, the five means for workplace diners into list **L2**, and the five means for the fast-foods into list **L3**. Then adjust the **WINDOW**. Next set **Plot1** for a time plot of **L1** and **L2**, and set **Plot2** for a time plot of **L1** and **L3**.

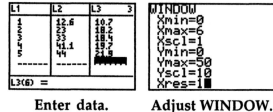

Enter data.   Adjust WINDOW.

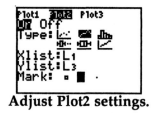

Adjust Plot1 settings.   Adjust Plot2 settings.

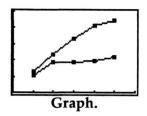

Graph.

We see that the patterns are not parallel; so it appears that we have an interaction.

13.2   Inference for Two-Way ANOVA

In this section, we give the **ANOVA2** program that can be used to perform two-way analysis of variance given the summary statistics having c observations per cell.

Example 13.11 The tables below give the summary statistics for the heart rate after six minutes of exercise on a treadmill. There were 200 subjects in each of the four combinations of male or female and athletic or sedentary. Use two-way ANOVA to test whether the mean heart rate is independent of sex and/or independent of lifestyle.

Means

	Runners	Control
Female	115.99	148.00
Male	103.98	130.00

Deviations

	Runners	Control
Female	15.97	16.27
Male	12.50	17.10

The ANOVA2 Program

```
PROGRAM:ANOVA2
:Disp "NO. OF ROWS"
:Input R
:Disp "NO. OF COL."
:Input S
:Disp "NO. PER CELL"
:Input C
:ClrList L₁,L₂,L₃,L₄,L₅
;For(I,1,R)
:I→L₁(I)
:sum(seq([A](I,J),J,1,S))/S→L₂(I)
:End
:sum(seq(L₂(I),I,1,R))/R→L₅(1)
:For(J,1,S)
:J→L₃(J)
:sum(seq([A](I,J),I,1,R))/R→L₄(J)
:End
:CS*sum(seq((L₂(I)-L₅(1))²,I,1,R))
 →L₅(2)
:CR*sum(seq((L₄(J)-L₅(1))²,J,1,S))
 →L₅(3)
:For(I,1,R)
:sum(seq(([A](I,J)+L₅(1)-L₂(I)-
 L₄(J))²,J,1,S))→L₆(I)
:End
:C*sum(seq(L₆(I),I,1,R))→L₅(4)
:If C=1
:Then
:(S-1)*L₅(2)/L₅(4)→U
:(R-1)*L₅(3)/L₅(4)→V
:Else
:For(I,1,R)
:sum(seq((C-1)[B](I,J)²,J,1,S))
 →L₆(I)
:End
:sum(seq(L₆(I),I,1,R))→L₅(5)
```

```
:(L₅(2)/(R-1))/(L₅(5)/(RS(C-1)))→U
:(L₅(3)/(S-1))/(L₅(5)/(RS(C-1)))→V
:(L₅(4)/(R-1)/(S-1))/(L₅(5)/
 (RS(C-1)))→W
:End
:R-1→M
:S-1→O
:(R-1)(S-1)→Z
:If C=1
:Then
:(R-1)(S-1)→N
:Else
:RS(C-1)→N
:End
:1-Fcdf(0,U,M,N)→T
:1-Fcdf(0,V,O,N)→P
:If C≠1
:Then
:1-Fcdf(0,W,Z,N)→Q
:End
:ClrHome
:Output(1,2,"ROW F")
:Output(1,8,round(U,5))
:Output(2,2,"ROW P")
:Output(2,8,round(T,8))
:Output(4,2,"COL F")
:Output(4,8,round(V,5))
:Output(5,2,"COL P")
:Output(5,8,round(P,8))
:If C≠1
:Then
:Output(7,2,"INT F")
:Output(7,8,round(W,5))
:Output(8,2,"INT P")
:Output(8,8,round(Q,8))
:End
```

To execute the program, enter the means into matrix **[A]** and the standard deviations into matrix **[B]**. (If there is only one observation per cell, then enter these values into matrix **[A]**.) When prompted, enter the numbers of rows, columns, and observations per cell. The program displays the F-statistics and P-values for the row variable, for the column variable, and for the interaction when there is more than one observation per cell.

Solution. First, enter the means into the 2 × 2 matrix [A] and enter the deviations into the 2 × 2 matrix [B]. Next, execute the **ANOVA2** program by entering the number of rows (2), the number of columns (2), and the number per cell (200).

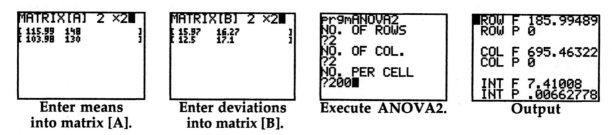

| Enter means into matrix [A]. | Enter deviations into matrix [B]. | Execute ANOVA2. | Output |

We obtain a *P*-value for each factor as well as for the interaction. The *P*-values for both the gender (row) and group (column) round to 0. If either the two genders or the two groups produced the same mean, then there would be virtually no chance of the measurements being as varied as they are. Thus, we can conclude that the means are significantly different.

The low *P*-value of 0.0066 for the interaction allows us to reject the hypothesis that there is no interaction among the four cells.

The **ANOVA2** program also stores the marginal means for the rows and columns in lists **L2** and **L4**. The overall mean is the first value 3.025 in list **L5**. The remaining entries in **L5** are the sum of squares for the row variable, column variable, interaction, and error.

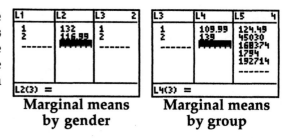

| Marginal means by gender | Marginal means by group |

The program enumerates lists **L1** and **L3** to allow for easy plotting of the marginal means. To graph, simply adjust the **WINDOW** settings, then set **Plot1** for a time plot of **L1** and **L2**, and set **Plot2** for a time plot of **L3** and **L4**.

  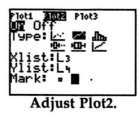 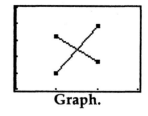

| Adjust WINDOW. | Adjust Plot1. | Adjust Plot2. | Graph. |

Exercise 13.22 The table below gives the amount of iron in certain foods, measured in milligrams of iron per 100 grams of cooked food, after samples of each food were cooked in each type of pot.

IRON	Food											
Type of pot	Meat				Legumes				Vegetables			
Aluminum	1.77	2.36	1.96	2.14	2.40	2.17	2.41	2.34	1.03	1.53	1.07	1.30
Clay	2.27	1.28	2.48	2.68	2.41	2.43	2.57	2.48	1.55	0.79	1.68	1.82
Iron	5.27	5.17	4.06	4.22	3.69	3.43	3.84	3.72	2.45	2.99	2.80	2.92

(a) Make a table giving the sample size, mean, and standard deviation for each type. (b) Plot the means. (c) Perform two-way ANOVA on the data regarding the main effects and interaction.

Solution. There are two factors, type of pot and type of food, each of which has three types. Thus, there are nine cells, each with four measurements per cell. To perform two-way ANOVA with the **ANOVA2** program, we first must compute the mean and standard deviation of each cell.

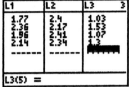

If we enter the four measurements per cell into lists **L1** through **L9**, then we can do the computations pairwise by successively entering the commands **2–Var Stats L1,L2**, then **2–Var Stats L3,L4**, and so on. The means and deviations for the first four cells are displayed below.

Enter data into lists.

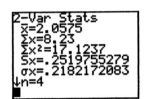 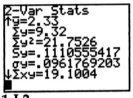 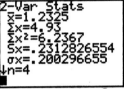 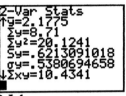

2–Var Stats L1,L2   **2–Var Stats L3,L4**

After doing all the computations, we obtain two new data tables.

Means	Meat	Legumes	Vegetables
Aluminum	2.0575	2.33	1.2325
Clay	2.1775	2.4725	1.46
Iron	4.68	3.67	2.79

Deviations	Meat	Legumes	Vegetables
Aluminum	0.2519755	0.1110555	0.2312826
Clay	0.6213091	0.0713559	0.4600724
Iron	0.6282781	0.1726268	0.2398611

Next, enter the means into matrix **[A]** and the standard deviations into matrix **[B]**. Then execute the **ANOVA2** program by entering the number of rows (3), the number of columns (3), and the number of measurements per cell (4).

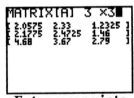 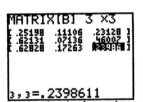 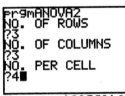

Enter means into matrix **[A]**.   Enter deviations into matrix **[B]**.   Execute ANOVA2.   Output

We obtain very low *P*-values for both factors and for the interaction. Therefore, we can conclude that, even with these small samples, there is a significant difference in the mean due to type of pot and due to type of food. Also, there is a statistically significant interaction.

After the **ANOVA2** program runs, the three marginal means for the rows (type of pot) are stored in list **L2**, and the three marginal means for the columns (food type) are stored in list **L4**. We now can plot either or both sets of marginal means.

L2	L3	L4	3
1.8733	1	2.9717	
2.0367	2	2.8242	
3.7133	3	1.8275	
------		------	

L3(4) =

Marginal means are stored in L2 and L4.

WINDOW
Xmin=0
Xmax=4
Xscl=1
Ymin=0
Ymax=4
Yscl=1
Xres=1■

Adjust WINDOW.

Plot1 Plot2 Plot3
On Off
Type: ⌐ ⌐ dln
     ⌐⊞ ⌐⊞ ⌐
Xlist:L1
Ylist:L2
Mark: ■ + ·

Adjust Plot1.

Plot1 Plot2 Plot3
On Off
Type: ⌐ ⌐ dln
     ⌐⊞ ⌐⊞ ⌐
Xlist:L3
Ylist:L4
Mark: □ ■ ·

Adjust Plot2.

Marginal means for type of pot

Marginal means for type of food

Both sets of marginal means

CHAPTER

14

```
{1,2,3,4,5,6,
      7,8,9,10}
{7,2,5,4,9,1}
  {6,3,10,8}
```

Bootstrap Methods and Permutation Tests

14.1 The Bootstrap Idea
14.2 First Steps in Using the Bootstrap
14.3 How Accurate Is a Bootstrap Distribution?
14.4 Bootstrap Confidence Intervals
14.5 Significance Testing Using Permutation Tests

Introduction

In this chapter, we demonstrate the bootstrap methods for estimating population parameters. Throughout, we use TI-83 Plus programs to perform the necessary resampling procedures.

14.1 The Bootstrap Idea

We first provide a TI-83 Plus program that performs resampling on a previously obtained random sample. Before executing the **BOOT** program, enter the random sample into list **L1**. Then bring up the program and enter the desired number of resamples. If you also want a confidence interval for the statistic, enter 1 for **CONF. INTERVAL?**; otherwise, enter 0. The program takes resamples from the entered random sample, enters their means into list **L2**, and displays the mean and bootstrap standard error of the resamples.

The BOOT Program

```
Program:BOOT
:dim(L₁)→N                    :randInt(1,N,N)→L₃
:Disp "NO. OF RESAMPLES"      :sum(seq(L₁(L₃(J)),J,1,N))/N→L₂(I)
:Input B                      :End
:Disp "CONF. INTERVAL?"       :ClrList L₃
:Input Y                      :1-Var Stats L₂
:If Y=1                       :ClrHome
:Then                         :Disp "AVG OF RESAMPLES"
:Disp "CONF. LEVEL"           :Disp x̄
:Input R                      :Disp "BOOT SE"
:"tcdf(0,X,N-1)"→Y₁           :Disp Sx
:solve(Y₁-R/2,X,2)→Q          :If Y=1
:1-Var Stats L₁               :Then
:x̄→X                          :Output(6,2,"CONF. INTERVAL")
:End                          :Output(7,2,X-Q*Sx)
:ClrList L₂                   :Output(8,2,X+Q*Sx)
:For(I,1,B)                   :End
```

Exercise 14.7 Here is an SRS of 20 guinea pig survival times (in days) during a medical trial. Create and inspect the bootstrap distribution of the sample mean for these data.

92	123	88	598	100	114	89	522	58	191
137	100	403	144	184	102	83	126	53	79

Solution. We will use the **BOOT** program with 100 resamples to create a bootstrap distribution. But first, we must enter the data into list **L1**. In this running of the program, we will not find a confidence interval, so we enter 0 for **CONF. INTERVAL?** when prompted. After taking a few minutes to execute, the program stores the 100 means from the resamples in list **L2**.

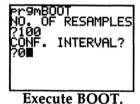
Enter data into L1.

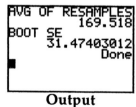
Execute BOOT.

Output

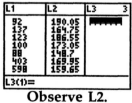
Observe L2.

To assess the normality of the distribution of the resample means, we shall make a histogram of the data in list **L2**. We adjust the **WINDOW** to reflect the range of these data, adjust the **STAT PLOT** settings, then press **GRAPH**. We see that the bootstrap distribution appears close to normal.

Adjust WINDOW.

Adjust STAT PLOT.

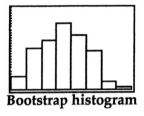
Bootstrap histogram

Exercise 14.8 Compute the standard error $s / \sqrt{n}$ for the 20 survival times in Exercise 14.7 and compare it with the bootstrap standard error from the resampling done in that exercise.

Solution. With the data entered into list **L1**, we can use the command **1–Var Stats L1** to compute the sample deviation s. Then retrieving the variable **Sx** from the **VARS Statistics** menu, we enter the command **Sx/√(20)** to compute the standard error.

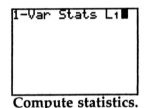
Compute statistics.

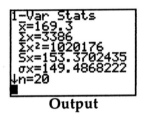
Output

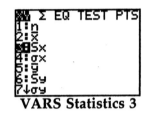
VARS Statistics 3

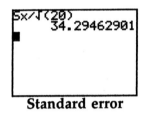
Standard error

We see that standard error of 34.294629 is slightly higher than the bootstrap standard error of 31.474 obtained from our 100 resamples in Exercise 14.7.

14.2   First Steps in Using the Bootstrap

In this section, we demonstrate how to compute a bootstrap t confidence interval for a population parameter. We begin though with a quick estimation of the bias obtained using the preceding exercises.

Exercise 14.9 Return to the 20 guinea pig survival times from Exercise 14.7.
(a) What is the bootstrap estimate of the bias?
(b) Give the 95% bootstrap t confidence interval for μ.
(c) Give the usual 95% one-sample t confidence interval.

Solution. (a) In Exercise 14.8, we found the sample mean of the data, along with the standard deviation, when using the **1–Var Stats** command. From the output display, we see that $\bar{x} = 169.3$. The average of all 100 resample means obtained in Exercise 14.7 was 169.518. Thus, the bootstrap estimate of bias is $169.518 - 169.3 = 0.218$.

(b) We now apply the formula $\bar{x} \pm t^* \mathrm{SE}_{boot}$, where the critical value t^* is from the $t(n-1)$ distribution and the sample size is $n = 20$. To find this critical value from the $t(19)$ distribution, we can use the **TSCORE** program (page 50). Doing so, we obtain $t^* = 2.093$.

The 95% bootstrap t confidence interval becomes
169.3 ± 2.093×31.474, or (103.425, 235.175).

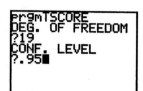 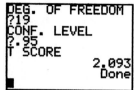

Note: This confidence interval would have been computed and displayed when executing the **BOOT** program in Exercise 14.7 had we entered **1** for **CONF. INTERVAL?**. A different bootstrap confidence interval is displayed below from another execution of the **BOOT** program.

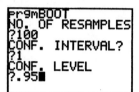 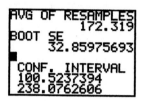

(c) To compute the standard one-sample t confidence interval, we use the **TInterval** feature from the **STAT TESTS** menu. After adjusting the **List** to **L1**, we calculate a 95% confidence interval of (97.521, 241.08).

 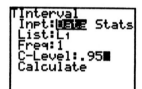 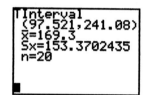

Confidence Interval for Trimmed Mean

Next, we demonstrate a program that will compute a bootstrap t confidence interval for a trimmed mean. Before executing the **BOOTTRIM** program given on the next page, enter the random sample into list **L1**. Then enter the desired number of resamples, the decimal amount to be trimmed at each end, and the desired confidence level when prompted. The program takes resamples from the entered random sample and enters their trimmed means into list **L2**. Then the trimmed mean of the original sample, the trimmed bootstrap standard error, and the confidence interval are displayed.

Example 14.5 The table below gives an SRS of 50 real estate sale prices in Seattle (in thousands of dollars) during 2002. Use the bootstrap t method to give a 95% confidence interval for the 25% trimmed mean sale price.

142	175	197.5	149.4	705	232	50	146.5	155	1850
132.5	215	116.7	244.9	290	200	260	449.9	66.407	164.95
362	307	266	166	375	244.95	210.95	265	296	335
335	1370	256	148.5	987.5	324.5	215.5	684.5	270	330
222	179.8	257	252.95	149.95	225	217	570	507	190

Solution. After entering the data into list **L1**, we simply execute the **BOOTTRIM** program. Here we use 50 resamples. The trimmed mean of the original sample is given as 240.562, and the desired confidence interval is (207.70224, 273.42176).

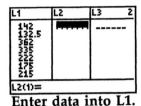
Enter data into L1.

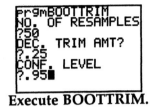
Execute BOOTTRIM.

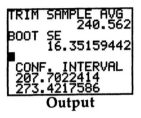
Output

Note: Due to programming differences, the trimmed mean given by the **BOOTTRIM** program may not agree with the value given by software. Because the sample size of 50 is not a multiple of 4, we cannot trim precisely 25% of the measurements from the high and low end. In the case of example 14.5 above, the **BOOTTRIM** program trimmed the lowest 12 and the highest 13 measurements before computing the trimmed sample mean.

The BOOTTRIM Program

```
Program:BOOTTRIM
:dim(L₁)→N
:Disp "NO. OF RESAMPLES"
:Input B
:Disp "DEC. TRIM AMT?
:Input M
:Disp "CONF. LEVEL"
:Input R
:"tcdf(0,X,N-1)"→Y₁
:solve(Y₁-R/2,X,2)→Q
:int(M*N)+1→L
:int((1-M)*N)→U
:SortA(L₁)
:sum(seq(L₁(I),I,L,U))/(U+1-L)→X
:ClrList L₂,L₄
:For(I,1,B)
:randInt(1,N,N)→L₃
:For(J,1,N)
:L₁(L₃(J))→L₄(J)
:End
:SortA(L₄)
:sum(seq(L₄(I),I,L,U))/(U+1-L)→L₂(I)
:End
:ClrList L₃,L₄
:1-Var Stats L₂
:ClrHome
:Disp "TRIM SAMPLE AVG"
:Disp X
:Disp "BOOT SE"
:Disp Sx
:Output(6,2,"CONF. INTERVAL")
:Output(7,2,X-Q*Sx)
:Output(8,2,X+Q*Sx)
```

Difference in Means

We conclude this section with a program that computes a bootstrap *t* confidence interval for the difference in means. Before executing the **BOOTPAIR** program, enter a random sample from the first population into list **L1** and enter a random sample from the second population into list **L2**. Then enter the desired number of resamples and the confidence level. The resampled differences in mean are stored in list **L3**. The difference of the original sample averages is displayed along with the bootstrap standard error and the confidence interval.

The BOOTPAIR Program

```
Program:BOOTPAIR
:dim(L₁)→N
:dim(L₂)→M
:Disp "NO. OF RESAMPLES"
:Input B
:Disp "CONF. LEVEL"
:Input R
:1-Var Stats L₁
:x̄→X
:1-Var Stats L₂
:x̄→Y
:ClrList L₃,L₄
:For(I,1,B)
:randInt(1,N,N)→L₅
:randInt(1,M,M)→L₆
:sum(seq(L₁(L₅(J)),J,1,N))/N→L₃(I)
:sum(seq(L₂(L₆(J)),J,1,M))/M→L₄(I)
:End
```

```
:1-Var Stats L₃
:Sx→S
:1-Var Stats L₄
:Sx→T
:int((S²/N+T²/M)²/((S²/N)²/
 (N-1)+(T²/M)²/(M-1)))→P
:√(S²/N+T²/M)→D
:"tcdf(0,X,P)"→Y₁
:solve(Y₁-R/2,X,2)→B
:B*D→E
:ClrHome
:Disp "DIFF OF AVGS"
:Disp X-Y
:Disp "BOOT SE"
:Disp D
:Output(6,2,"CONF. INTERVAL")
:Output(7,2,X-Y-E)
:Output(8,2,X-Y+E)
```

Exercise 14.14 Following are the scores from Table 14.3 on a test of reading ability for two groups of third-grade students.

Treatment group				Control group			
24	61	59	46	42	33	46	37
43	44	52	43	43	41	10	42
58	67	62	57	55	19	17	55
71	49	54		26	54	60	28
43	53	57		62	20	53	48
49	56	33		37	85	42	

(a) Bootstrap the difference in means $\bar{x}_1 - \bar{x}_2$ and give the bootstrap standard error and a 95% bootstrap t confidence interval.
(b) Compare the bootstrap results with a two-sample t confidence interval.

Solution. (a) We first enter the treatment scores into list **L1** and the control scores into list **L2**. We shall use 30 resamples to bootstrap the difference in means and to obtain a 95% confidence interval. Upon executing the **BOOTPAIR** program, we obtain a bootstrap standard error of about 1.074 and a confidence interval of (7.769, 12.1399).

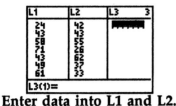
Enter data into L1 and L2.

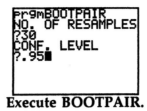
Execute BOOTPAIR.

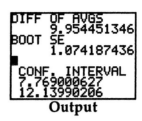
Output

(b) To find a traditional two-sample t confidence interval, we can use the **2–SampTInt** feature from the **STAT TESTS** menu. Because we are not assuming normal populations with the same variance, we do not pool the sample variances. Upon calculating, we obtain a 95% confidence interval of $(1.233, 18.676)$. This interval is much wider than the bootstrap interval due to the much larger standard error obtained when using the original sample deviations s_x and s_y as opposed to the sample deviations from the collection of resample means.

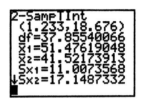

14.3   How Accurate Is a Bootstrap Distribution?

Below we work an exercise (with some slight variations) to demonstrate how the bootstrap distribution varies with the sample size.

Exercise 14.23 (a) Draw an SRS of size 10 from a $N(8.4, 14.7)$ population. What is the exact distribution of the sampling mean $\bar{x}$ for this sample size? (b) Bootstrap the sample mean using 100 resamples. Give a histogram of the bootstrap distribution and the bootstrap standard error. (c) Repeat the process for a sample of size 40.

Solution. (a) For an SRS of size $n = 10$ from a $N(8.4, 14.7)$ population, $\bar{x} \sim N(8.4, 14.7/\sqrt{10}) \approx N(8.4, 4.648548)$. To draw such an SRS, we will use the **randNorm(** command from the **MATH PRB** menu. Enter the command **randNorm(8.4, 24.7, 10)$\rightarrow$L1** to store the SRS in list **L1**.

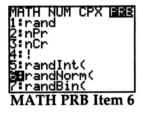
MATH PRB Item 6

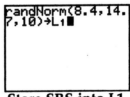
Store SRS into L1.

Observe L1.

(b) With the sample in list **L1**, we can execute the **BOOT** program to bootstrap the sample mean. In this case, we enter **0** for **CONF. INTERVAL?** because we are only interested here in the bootstrap distribution. Due to the small sample size, we obtain statistics of this bootstrap distribution that differ noticeably from those of the $N(8.4, 4.648548)$ distribution. A histogram of the resample means stored in list **L2** is also shown on the following page.

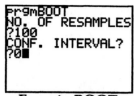
Execute BOOT.

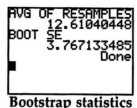
Bootstrap statistics

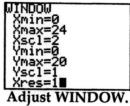
Adjust WINDOW.

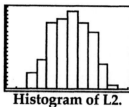
Histogram of L2.

(c) We now repeat the process for an SRS of size 40. With this sample size, the bootstrap distribution statistics become much closer to the actual $N(8.4, 14.7/\sqrt{40}) = N(8.4, 2.324274)$ distribution of $\bar{x}$.

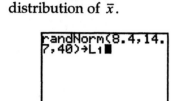
Generate new SRS.

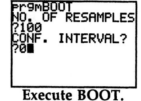
Execute BOOT.

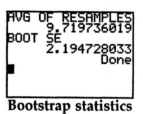
Bootstrap statistics

14.4 Bootstrap Confidence Intervals

In this section, we demonstrate how to find a bootstrap percentile confidence interval. We also provide another program that will bootstrap the correlation coefficient or the regression slope for two related variables.

Exercise 14.28 For the data given below, calculate the 95% one-sample t confidence interval for this sample. Then give a 95% bootstrap percentile confidence interval for the mean.

109	123	118	99	121	134	126	114	129	123	171	124	111	125	128
154	121	123	118	106	108	112	103	125	137	121	102	135	109	115
125	132	134	126	116	105	133	111	112	118	117	105	107		

Solution. To find the traditional confidence interval, we will use the **TInterval** feature from the **STAT TESTS** menu. We enter the data into list **L1**, adjust the **Inpt**, **List**, and **C–Level**, and calculate. We obtain a 95% confidence interval of (116.35, 124.81), or 120.58 ± 4.23.

Enter data into L1.

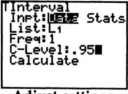
STAT TESTS item 8

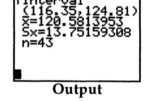
Adjust settings.

Output

Next we will execute the **BOOT** program with 80 resamples. Here we enter 0 for **CONF. INTERVAL?** because we will be finding a percentile confidence interval rather than a bootstrap t confidence interval. After the program executes, the resample means are stored in list **L2**. By sorting this list, we can find a bootstrap percentile confidence interval.

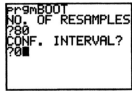

| Execute BOOT. | Sort list L2. | Third element in L2 | 78th element in L2 |

After the **BOOT** program completes, we enter the command **SortA(L2** to place the resample means into increasing order. Because we want a 95% percentile interval, we now eliminate the upper 2.5% and lower 2.5% from list **L2**. And because $0.025 \times 80 = 2$, we use the third and 78th elements of the sorted list as the endpoints of our percentile confidence interval. By observing these values in the sorted **L2** list, we find our desired interval to be (116.86, 123.3). Using $\bar{x} = 120.58$ from the original sample in list **L1**, this interval can also be written as $(120.58 - 3.72, 120.58 + 2.72)$.

The BOOTCORR Program

```
Program:BOOTCORR
:dim(L₁)→N
:ClrList L₃,L₄,L₅
:Disp "1=CORRELATION","2=REG. SLOPE"
:Input W
:Disp "NO. OF RESAMPLES"
:Input B
:Disp "CONF. LEVEL"
:Input R
:For(I,1,B)
:randInt(1,N,N)→L₆
:For(J,1,N)
:L₁(L₆(J))→L₄(J)
:L₂(L₆(J))→L₅(J)
:End
:LinReg(ax+b) L₄,L₅
:If W=1
:Then
:r→L₃(I)
:Else
:a→L₃(I)
:End
:End
:LinReg(ax+b) L₁,L₂
:If W=1
:Then
:r→X

:Else
:a→X
:End
:ClrList L₄,L₅,L₆
:1-Var Stats L₃
:"tcdf(0,X,N-1)"→Y₁
:solve(Y₁-R/2,X,2)→Q
:ClrHome
:If W=1
:Then
:Disp "SAMPLE CORR."
:Else
:Disp "REG. SLOPE"
:End
:Disp X
:Disp "BOOT SE"
:Disp Sx
:Output(6,2,"CONF. INTERVAL")
:If W=1
:Then
:Output(7,2,max(⁻1,X-Q*Sx))
:Output(8,2,min(X+Q*Sx,1))
:Else
:Output(7,2,X-Q*Sx)
:Output(8,2,X+Q*Sx)
:End
```

The **BOOTCORR** program can be used to bootstrap the correlation coefficient or the regression slope for paired sample data. Before executing the program, enter the random sample data into lists **L1** and **L2**. When prompted, enter **1** to bootstrap the correlation coefficient or enter **2** to bootstrap the regression slope. Then enter the desired number of resamples and the confidence level. The resampled statistics are stored in list **L3**, and the statistic of the original sample data is displayed along with the bootstrap standard error and the confidence interval.

Exercise 14.43 The following table gives the 2002 salaries and career batting averages for 50 randomly selected MLB players (excluding pitchers).

Player	Salary	Avg	Player	Salary	Avg	Player	Salary	Avg
1.	$9,500,000	0.269	18.	$3,450,000	0.242	35.	$630,000	0.324
2.	$8,000,000	0.282	19.	$3,150,000	0.273	36.	$600,000	0.200
3.	$7,333,333	0.327	20.	$3,000,000	0.250	37.	$500,000	0.214
4.	$7,250,000	0.259	21.	$2,500,000	0.208	38.	$325,000	0.262
5.	$7,166,667	0.240	22.	$2,400,000	0.306	39.	$320,000	0.207
6.	$7,086,668	0.270	23.	$2,250,000	0.235	40.	$305,000	0.233
7.	$6,375,000	0.253	24.	$2,125,000	0.277	41.	$285,000	0.259
8.	$6,250,000	0.238	25.	$2,100,000	0.227	42.	$232,500	0.250
9.	$6,200,000	0.300	26.	$1,800,000	0.307	43.	$227,500	0.278
10.	$6,000,000	0.247	27.	$1,500,000	0.276	44.	$221,000	0.237
11.	$5,825,000	0.213	28.	$1,087,500	0.216	45.	$220,650	0.235
12.	$5,625,000	0.238	29.	$1,000,000	0.289	46.	$220,000	0.243
13.	$5,000,000	0.245	30.	$950,000	0.237	47.	$217,500	0.297
14.	$4,900,000	0.276	31.	$800,000	0.202	48.	$202,000	0.333
15.	$4,500,000	0.268	32.	$750,000	0.344	49.	$202,000	0.301
16.	$4,000,000	0.221	33.	$720,000	0.185	50.	$200,000	0.224
17.	$3,625,000	0.301	34.	$675,000	0.234			

(a) Calculate the sample correlation between salary and average.
(b) Bootstrap the correlation and give a 95% confidence interval for the correlation.
(c) Calculate the least-squares regression line to predict average from salary. Give the traditional 95% t confidence interval for the slope of the regression line.
(d) Bootstrap the regression model. Give a 95% bootstrap t confidence interval and a 95% bootstrap percentile confidence interval for the regression slope.

Solution. (a) and (c) We can find the sample correlation, least-squares regression line, and traditional confidence interval for the slope using techniques from Chapter 10. First, we enter the data into lists, say lists **L1** and **L2**. Next, we use the **LinReg(a+bx)** command (item 8 from the **STAT CALC** menu) and enter the command **LinReg(a+bx) L1,L2**. We obtain a sample correlation of $r \approx 0.1067575$. The regression line is given as $y = 0.25291 + 1.4769 \times 10^{-9} x$. Thus, we see that the slope appears to be 0.

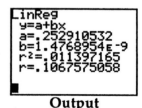

Enter data in L1, L2.   STAT CALC item 8   Enter the command.   Output

We can find a confidence interval for the slope using the **REG1** program. After running the program, we find a 95% t confidence interval for the slope of the regression line to be $(-2.51496 \times 10^{-9}, 5.46875 \times 10^{-9})$.

 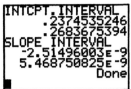

(b) and (d) We now execute the **BOOTCORR** program twice using 40 resamples each time. In the first running, we enter **1** to bootstrap the correlation. In the second running, we enter **2** to bootstrap the regression slope.

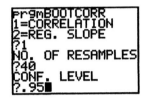 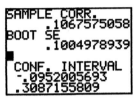 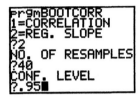 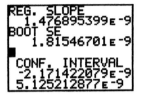

In the first running, we obtain a 95% bootstrap interval for the correlation of (−0.0952, 0.3087). In the second running, we obtain a 95% bootstrap interval for the regression slope of $(-2.17142 \times 10^{-9}, 5.12521 \times 10^{-9})$.

The 40 resample regression slopes are stored in list **L3** after option 2 of the **BOOTCORR** program ends. So next, we enter the command **SortA(L3)** to place the list increasing order. Because $0.025 \times 40 = 1$, the endpoints of the 95% percentile confidence interval are the second and 39th elements of the sorted list. The interval here is $(-2.0174 \times 10^{-9}, 4.74556 \times 10^{-9})$.

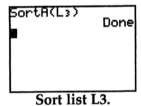

Sort list L3.

L1	L2	L3	3
9.5E6	.269	-3E-9	
8E6	.282	-2E-9	
7.33E6	.327	-2E-9	
7.25E6	.259	-1E-10	
7.17E6	.24	-5E-12	
7.09E6	.27	1E-10	
6.38E6	.253	2E-10	

L3(2) = -2.0174031...

Second element in L3

L1	L2	L3	3
630000	.324	3.9E-9	
600000	.2	4E-9	
500000	.214	4.1E-9	
325000	.262	4.1E-9	
320000	.207		
305000	.233	6E-9	
285000	.259	------	

L3(39)=4.74556519...

39th element in L3

14.5 Significance Testing Using Permutation Tests

We conclude this chapter with two programs that will simulate some of the bootstrap permutation tests. The **BOOTTEST** program performs a permutation test for the difference in means and the **BTPRTEST** program performs a permutation test for either the difference in paired means or for the correlation. We note that for large samples and many resamples, the programs will take several minutes to execute.

Before executing the **BOOTTEST** program, enter data from the first population into list **L1** and enter data from the second population into list **L2**. When prompted, enter 1, 2, or 3 to designate the desired alternative $\mu_1 < \mu_2$, $\mu_1 > \mu_2$, or $\mu_1 \neq \mu_2$. The resampled differences in permuted mean are ordered and then stored in list **L3**. The program displays the difference in the original sample means $\bar{x}_1 - \bar{x}_2$ and the P-value.

The BOOTTEST Program

```
Program:BOOTTEST
:Disp "1 = ALT. <","2 = ALT. >","3 = ALT. ≠"
:Input C
:Disp "NO. OF RESAMPLES"
:Input B
:dim(L₁)→N
:dim(L₂)→M
:augment(L₁,L₂)→L₁
:ClrList L₃
:For(S,1,B)
:ClrList L₆
:seq(J,J,1,N+M)→L₄
:For(I,1,M)
:ClrList L₅
:randInt(1,N+M-I+1)→A
:L₄(A)→L₆(I)
:1→K
:While K<A
:L₄(K)→L₅(K)
:1+K→K
:End
:A→K
:While K≤(N+M-I)
:L₄(K+1)→L₅(K)
:1+K→K
:End
:L₅→L₄
:End
:sum(seq(L₁(L₅(J)),J,1,N))/N→X
:sum(seq(L₁(L₆(J)),J,1,M))/M→Y
:X-Y→L₃(S)
:End

:ClrList L₄
:seq(L₁(I),I,1,N)→L₁
:mean(L₁)→X
:mean(L₂)→Y
:SortA(L₃)
:X-Y+1→L₃(B+1)
:0→K
:While L₃(K+1)<(X-Y)
:K+1→K
:End
:K/B→P
:While L₃(K+1)≤(X-Y)
:K+1→K
:End
:(B-K)/B→Q
:seq(L₃(I),I,1,B)→L₃
:ClrHome
:Disp "DIFF IN MEANS"
:Disp X-Y
:Disp "P VALUE"
:If C=1
:Then
:Disp P
:Else
:If C=2
:Then
:Disp Q
:Else
:Disp 2*min(P,Q)
:End
:End
```

Exercise 14.48 A fast food restaurant customer complains that people 60 years old or older are given fewer French fries than people under 60. The owner responds by gathering data without knowledge of the employees. Below are the data on the weight of French fries (in grams) from random samples of the two groups of customers. Perform a permutation test using an appropriate alternative hypothesis and give the P-value.

Age < 60:	75	77	80	69	73	76	78	74	75	81
Age ≥ 60:	68	74	77	71	73	75	80	77	78	72

Solution. We let μ_1 be the average weight of French fries served to customers under age 60, and let μ_2 be the average weight served to customers age 60 or older. We shall test H_0: $\mu_1 = \mu_2$ with the alternative H_a: $\mu_1 > \mu_2$.

First, we enter the sample weights for the under age 60 customers into list **L1** and the weights for the other group into list **L2**. Then we execute the **BOOTTEST** program by entering **2** to designate the alternative $\mu_1 > \mu_2$. Below are the results from 50 resamples.

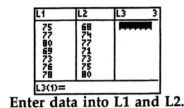 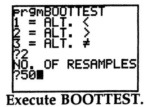 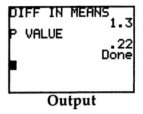

Enter data into L1 and L2. Execute BOOTTEST. Output

The 50 resampled differences in permuted mean are stored in list **L3**. The difference in the original sample means is $\bar{x}_1 - \bar{x}_2 = 1.3$ and 22% of the means in L3 are greater than 1.3. Based on this sample of permutations, if $\mu_1 = \mu_2$, then there would be about a 22% chance of $\bar{x}_1 - \bar{x}_2$ being as high as 1.3 with samples of these sizes. This P-value of 0.22 does not provide enough evidence to reject H_0.

Using the **2–SampTTest** feature, we see that the traditional two-sample t test gives a P-value of 0.21243 and leads us to the same conclusion.

 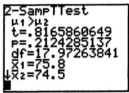

Permutation Test for Paired Data

When we have paired sample data, such as "before and after" measurements, then we often consider the differences in measurements as one sample. The hypothesis test H_0: $\mu_1 = \mu_2$ then becomes H_0: $\mu_1 - \mu_2 = 0$ and may be tested with this one sample using a traditional t test or with many resamples using the permutation test. The **BTPRTEST** program may be used for this permutation test. It also may be used to test whether the correlation equals 0.

Before executing the **BTPRTEST** program, enter the data into lists **L1** and **L2**. When prompted, enter **1** or **2** to designate the enter desired test, then enter **1, 2,** or **3** to designate the desired alternative. The resampled permuted pair differences in mean (or correlation) are ordered and stored in list **L3**. The statistic from the original paired sample is displayed along with the P-value of the permutation test.

Exercise 14.64 Below is the data from Exercise 7.27 that gives the total body bone mineral content of eight subjects as measured by two different X-ray machine operators. Perform a matched pairs permutation test whether the two operators have the same mean.

	Subject							
Operator	1	2	3	4	5	6	7	8
1	1.328	1.342	1.075	1.228	0.939	1.004	1.178	1.286
2	1.323	1.322	1.073	1.233	0.934	1.019	1.184	1.304

The BTPRTEST Program

```
Program:BTPRTEST
:Disp "1 = PAIRED MEAN","2 = CORRELATION"
:Input T
:ClrHome
:Disp "1 = ALT. <","2 = ALT. >","3 = ALT. ≠"
:Input C
:Disp "NO. OF RESAMPLES"
:Input B
:dim(L₁)→N
:ClrList L₃
:For(I,1,B)
:ClrList L₅,L₆
:If T=2
:Then
:L₁→L₅
:randInt(1,N,N)→L₄
:For(J,1,N)
:L₂(L₄(J))→L₆(J)
:End
:LinReg(ax+b) L₅,L₆
:r→L₃(I)
:Else
:For(J,1,N)
:randInt(0,1)→A
:If A=0
:Then
:L₁(J)→L₅(J)
:L₂(J)→L₆(J)
:Else
:L₁(J)→L₆(J)
:L₂(J)→L₅(J)
:End
:End
:If T=1
:Then
:mean(L₅)-mean(L₆)→L₃(I)
:End
:End
:If T=1
```

```
:Then
:mean(L₁)-mean(L₂)→X
:Else
:LinReg(ax+b) L₁,L₂
:r→X
:End
:SortA(L₃)
:X+1→L₃(B+1)
:0→K
:While L₃(K+1)<X
:K+1→K
:End
:K/B→P
:While L₃(K+1)≤X
:K+1→K
:End
:(B-K)/B→Q
:seq(L₃(I),I,1,B)→L₃
:ClrHome
:If T=1
:Then
:Disp "DIFF. IN MEANS"
:Else
:Disp "SAMPLE CORR."
:End
:Disp X
:Disp "P VALUE"
:If C=1
:Then
:Disp P
:Else
:If C=2
:Then
:Disp Q
:Else
:Disp 2*min(P,Q)
:End
:End
```

Solution. We let μ_1 be the average measurement from Operator 1 and let μ_2 be the average measurement from Operator 2. We shall test $H_0: \mu_1 = \mu_2$ with the alternative $H_a: \mu_1 \neq \mu_2$.

First, we enter the measurements into lists **L1** and **L2**. Then we execute the **BTPRTEST** program by entering **1** to designate a paired mean, and then entering **3** for the alternative $\mu_1 \neq \mu_2$. Next, we display the results from 50 resamples.

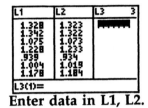 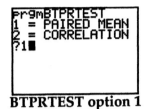 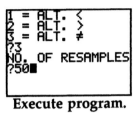 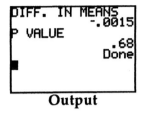

Enter data in L1, L2. BTPRTEST option 1 Execute program. Output

The 50 resampled permuted pair differences in mean are stored in list **L3**. The difference in the original sample means is $\bar{x}_1 - \bar{x}_2 = -0.0015$ and $68\%/2 = 34\%$ of the means in **L3** are less than -0.0015. This large P-value means that we do not have significant evidence to argue that there is a difference in the operator's mean measurement.

Exercise 14.54 Use the data from Table 14.3, given previously in Exercise 14.43, to test whether the correlation between salary and batting average is greater than 0.

Solution. First, we enter the data into lists **L1** and **L2**. Then to test the hypothesis $H_0: \rho = 0$ with alternative $H_a: \rho > 0$, we execute option 2 of the **BTPRTEST** program using the second alternative with 50 resamples.

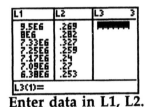 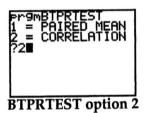 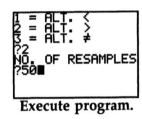 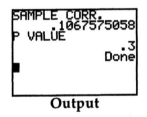

Enter data in L1, L2. BTPRTEST option 2 Execute program. Output

Upon running the program, the 50 resample correlations are stored in list **L3**. The sample correlation is $r \approx 0.1067575$. In this case, 15 of the resample correlations were greater than r, which gives us a one-sided P-value of 0.30. Thus, we do not have significant evidence to reject H_0 with this sample based upon 50 resamples. Our P-value of 0.30 compares favorably with the P-value of 0.23 obtained with the traditional linear regression t test.

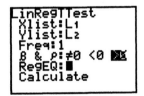 

CHAPTER
15

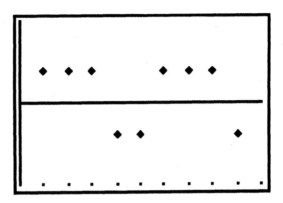

Nonparametric
Tests

Introduction

In this chapter, we provide some supplementary programs for performing several nonparametric hypothesis tests.

15.1   The Wilcoxon Rank Sum Test

We first provide the program **RANKSUM** to perform the Wilcoxon rank sum test on data from two populations. To execute the program, we must enter the data into lists **L1** and **L2**. The program sorts each list, then merges and sorts the lists into list **L3**. Then it stores the rank of each measurement in **L3** next to it in list **L4**. All sequences of ties are assigned an average rank.

The Wilcoxon test statistic W is the sum of the ranks from **L1**. Assuming that the two populations have the same continuous distribution (and no ties occur), then W has a mean and standard deviation given by

$$\mu = \frac{n_1(N+1)}{2} \quad \text{and} \quad \sigma = \sqrt{\frac{n_1 n_2(N+1)}{12}}$$

where n_1 is the sample size from **L1**, n_2 is the sample size from **L2**, and $N = n_1 + n_2$.

We test the null hypothesis H_0: no difference in distributions. A one-sided alternative is H_a: the first population yields higher measurements. We use this alternative if we expect or see that W is much higher sum than its expected sum of ranks μ. In this case, the P-value is given by a normal approximation. We let $X \sim N(\mu, \sigma)$ and compute the right-tail $P(X \geq W)$ (using continuity correction if W is an integer).

If we expect or see that W is the much lower sum than its expected sum of ranks μ, then we should use the alternative H_a: first population yields lower measurements. In this case, the P-value is given by the left-tail $P(X \leq W)$, again using continuity correction if needed.

If the two sums of ranks are close, then we could use a two-sided alternative H_a: there is a difference in distributions. In this case, the P-value is given by twice the smallest tail value: $2P(X \geq W)$ if $W > \mu$, or $2P(X \leq W)$ if $W < \mu$.

The **RANKSUM** program displays the expected sum of ranks from the first list and the actual sums of the ranks from **L1** and **L2**. It also displays the smallest tail value created by the test statistic. That is, it displays $P(X \geq W)$ if $W > \mu$, and it displays $P(X \leq W)$ if $W < \mu$. Conclusions for any alternative then can be drawn from this value. We note that if there are ties, then the validity of this test is questionable.

Exercise 15.1 Below are language usage scores of kindergarten students who were classified as high progress readers or low progress readers. Is there evidence that the scores of high progress readers are higher than those of low progress readers on Story 1? Carry out a two-sample t test. Then carry out the Wilcoxon rank sum test and compare the conclusions for each test.

Child	Progress	Story 1 score	Story 2 score
1	high	0.55	0.80
2	high	0.57	0.82
3	high	0.72	0.54
4	high	0.70	0.79
5	high	0.84	0.89
6	low	0.40	0.77
7	low	0.72	0.49
8	low	0.00	0.66
9	low	0.36	0.28
10	low	0.55	0.38

The RANKSUM Program

```
PROGRAM:RANKSUM
:ClrList L₄
:SortA(L₁)
:SortA(L₂)
:augment(L₁,L₂)→L₃
:SortA(L₃)
:dim(L₁)→M
:dim(L₂)→N
:L₃→L₆
:L₃(1)-1→L₃(M+N+1)
:1→B
:1→I
:Lbl 1
:While I<(M+N)
:If L₃(I)<L₃(I+1)
:Then
:B→L₄(I)
:1+I→I
:1+B→B
:Goto 1
:Else
:1→J
:B→S
:Lbl 2
:While L₃(I)=L₃(I+J)
:S+B+J→S
:1+J→J
:Goto 2
:End
:S/J→T
:For(K,0,J-1)
:T→L₄(I+K)
:End
:I+J→I
:B+J→B
:Goto 1
:End
```

```
:End
:If I=M+N
:Then
:M+N→L₄(I)
:End
:1→I
:0→S
:0→J
:Lbl 3
:While I≤M
:Lbl 4
:If L₁(I)=L₃(I+J)
:Then
:S+L₄(I+J)→S
:Else
:1+J→J
:Goto 4
:End
:I+1→I
:Goto 3
:End
:L₆→L₃
:ClrList L₆
:(M+N)(M+N+1)/2→R
:M*(M+N+1)/2→U
:√(M*N*(M+N+1)/12)→D
:(abs(S-U)-.5)/D→Z
:If int(S)≠S:(abs(S-U))/D→Z:End
:0.50-normalcdf(0,Z,0,1)→P
:Disp "EXPECTED 1ST SUM"
:Disp .5*M*(M+N+1)
:Disp "SUMS OF RANKS"
:Disp {S,R-S}
:If S=U:Disp "NO DIFFERENCE"
:If S<U:Disp "LEFT TAIL",round(P,4)
:If S>U:Disp "RIGHT TAIL",round(P,4)
```

Solution. First, we enter the five low-progress Story 1 scores into **L1** and the five high-progress Story 1 scores into **L2**. Then we apply the **2–SampTTest** feature (item 4 from the **STAT TESTS** menu) to test the hypothesis $H_0 : \mu_1 = \mu_2$ versus the alternative $H_a : \mu_1 < \mu_2$.

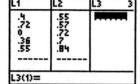 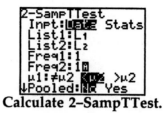 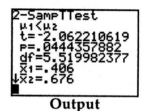

Enter data into L1 and L2.   Calculate 2–SampTTest.   Output

We obtain a t statistic of -2.06221 and a P-value of 0.044436 for the one-sided alternative. With the rather small P-value, we have significant evidence to reject H_0 and say that the average score of all high-progress readers is higher than the average score of all low-progress readers on Story 1. For if H_0 were true, then there would be only a 0.044436 probability of obtaining a high-progress sample mean that is so much larger than the low-progress sample mean (0.676 compared to 0.406).

Now for the Wilcoxon rank sum test, we use H_0: same distribution for both groups versus H_a: high-progress readers score higher on Story 1.

The Wilcoxon test statistic is the sum of ranks from **L1** in which we entered the low-progress scores. With the data entered into lists **L1** and **L2**, we now execute the **RANKSUM** program, then observe the sorted data and ranks in lists **L3** and **L4**.

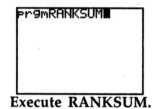 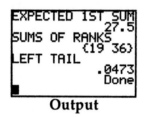  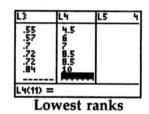

Execute RANKSUM.   Output   Highest ranks   Lowest ranks

The sum of the ranks from the low-progress readers is 19, which is lower than the expected average of $\mu = (5 \times 11)/2 = 27.5$. According to the Wilcoxon test, if the distributions were the same, then there would be only be a 0.0473 probability (from the left-tail value) of the low-progress sum of ranks being so much smaller than the expected average of 27.5. Therefore, we should reject H_0 in favor of the alternative.

In this case, the Wilcoxon P-value is slightly higher than the t test P-value; however, both are low enough to result in the same conclusion.

Exercise 15.7 Below is a comparison of the number of tree species in unlogged plots in the rain forest of Borneo with the number of species in plots logged eight years earlier.

Unlogged	22	18	22	20	15	21	13	13	19	13	19	15
Logged	17	4	18	14	18	15	15	10	12			

Does logging significantly reduce the mean number of species in a plot after eight years? State the hypotheses, do a Wilcoxon rank sum test, and state your conclusion.

Solution. We will test the hypothesis H_0: no difference in median versus the alternative H_a: unlogged median is higher. To do so, we first enter the unlogged measurements into list **L1** and enter the logged measurements into list **L2**. Then we execute the **RANKSUM** program which produces the following results:

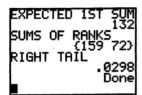 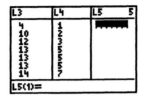 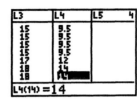

We note that there are 21 measurements with 12 unlogged measurements. If there were no difference in median, then we would expect the sum of ranks from **L1** to be $\mu = (12 \times 22)/2 = 132$. But if there were no difference in median, then there would be only a 2.98% chance of the sum of ranks from **L1** being as high as 159. This low P-value gives significant evidence to reject H_0 in favor of the alternative.

15.2 The Wilcoxon Signed Rank Test

Here we provide the **SIGNRANK** program to perform the Wilcoxon signed rank test on data sets of size n from two populations. To execute the program, we must enter the data into lists **L1** and **L2**. The program will sort the absolute value of the differences **L2 − L1** into list **L3**, but it will disregard any zero differences. The population size n is decreased so as to count only the non-zero differences. Then the program puts the rank of each measurement in **L3** next to it in **L4**. All sequences of ties are assigned an average rank.

The Wilcoxon test statistic W is the sum of the ranks from the positive differences. Assuming that the two populations have the same continuous distribution (and no ties occur), then W has a mean and standard deviation given by

$$\mu = \frac{n(n+1)}{4} \quad \text{and} \quad \sigma = \sqrt{\frac{n(n+1)(2n+1)}{24}}$$

We test the null hypothesis H_0: no difference in distributions. A one-sided alternative is H_a: second population yields higher measurements. We use this alternative if we expect or see that W is a much higher sum, which means that there were more positive differences in **L2 − L1**. In this case, the P-value is given by a normal approximation. We let $X \sim N(\mu, \sigma)$ and compute the right-tail $P(X \ge W)$ (using continuity correction if W is an integer).

If we expect or see that W is the much lower sum, then there were more negative differences. Now we should use the alternative H_a: second population yields lower measurements. In this case, the P-value is given by the left-tail $P(X \le W)$, again using continuity correction if needed.

If the two sums of ranks are close, we could use a two-sided alternative H_a: there is a difference in distributions. In this case, the P-value is given by twice the smallest tail value: $2P(X \ge W)$ if $W > \mu$, or $2P(X \le W)$ if $W < \mu$.

The SIGNRANK Program

```
PROGRAM:SIGNRANK
:ClrList L₃,L₄,L₅
:0→L₄(1):0→L₅(1):0→L₃(1)
:0→S:0→R:0→U:0→V:1→J
:For(I,1,dim(L₁),1)
:If L₂(I)-L₁(I)≠0
:Then
:abs(L₂(I)-L₁(I))→L₃(J)
:1+J→J
:End:End
:If L₃(1)>0
:Then
:SortA(L₃):dim(L₃)→N
:L₃→L₆:L₃(1)-1→L₃(N+1)
:1→B:1→I
:Lbl 1
:If I<N
:Then
:If L₃(I)<L₃(I+1)
:Then
:B→L₄(I)
:1+I→I:1+B→B
:Goto 1
:Else
:1→J:B→S
:Lbl 2
:If L₃(I)=L₃(I+J)
:Then
:S+B+J→S:1+J→J
:Goto 2
:End
:S/J→T
:For(K,0,J-1)
:T→L₄(I+K)
:End
:I+J→I:B+J→B
:End
:Goto 1
:End
```

```
:If I=N
:Then
:N→L₄(I)
:End
:0→J
:For(I,1,dim(L₁))
:If (L₂(I)-L₁(I))>0
:Then
:1+J→J:abs((L₂(I)-L₁(I))→L₅(J)
:End:End
:SortA(L₅)
:1→I:0→J
:If L₅(1)>0
:Then
:Lbl 3
:While I≤dim(L₅)
:Lbl 4
:If L₅(I)=L₃(I+J)
:Then
:V+L₄(I+J)→V
:Else
:1+J→J
:Goto 4
:End
:I+1→I
:Goto 3
:End:End
:L₆→L₃:ClrList L₆
:N(N+1)/2→R:N(N+1)/4→U
:√(N(N+1)(2N+1)/24)→D
:(abs(V-U)-.5)/D→Z
:If int(V)≠V:(abs(V-U))/D→Z:End
:0.50-normalcdf(0,Z,0,1)→P
:Disp "EXP. + SUM"
:Disp N*(N+1)/4
:Disp "SUMS -,+ RANKS"
:Disp {R-V,V}
:If V=U:Disp "NO DIFFERENCE"
:If V<U:Disp "LEFT TAIL",round(P,4)
:If V>U:Disp "RIGHT TAIL",round(P,4)
```

The **SIGNRANK** program displays the sums of the ranks of the negative differences and of the positive differences as well as the smallest tail value created by the test statistic. That is, it displays $P(X \geq W)$ if $W > \mu$, or $P(X \leq W)$ if $W < \mu$. Conclusions for any alternative can then be drawn from this value. Again, we note that if there are ties, then the validity of this test is questionable.

Exercises 15.13 Here are data for heart rates for five subjects and two treatments. Use the Wilcoxon signed rank procedure to reach a conclusion about the effect of the language institute. Show the assignment of ranks in the calculation of the test statistic.

	Low rate		Medium rate	
Subject	Resting	Final	Resting	Final
1	60	75	63	84
2	90	99	69	93
3	87	93	81	96
4	78	87	75	90
5	84	84	90	108

Does exercise at the low rate raise heart rate significantly? State hypotheses in terms of the median increase in heart rate and apply the Wilcoxon signed rank test.

Solution. We will test the hypothesis H_0: For the low rate, resting and final heart rates have the same median versus H_a: final heart rates are higher.

Enter the five low rate resting heart rates into list **L1** and the five low rate final heart rates into list **L2**. The alternative means that there should be more positive differences, so that the sum of the positive ranks should be higher. Therefore, the P-value comes from the right-tail probability created by the test statistic. After the data is entered, execute the **SIGNRANK** program.

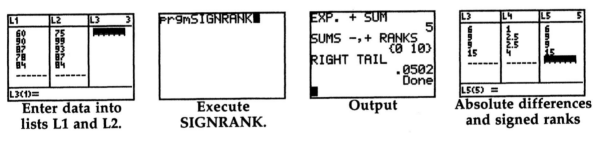

| Enter data into lists L1 and L2. | Execute SIGNRANK. | Output | Absolute differences and signed ranks |

List **L3** now contains the ordered absolute values of the four non-zero differences. Their corresponding (averaged) ranks are adjacent in list **L4**. List **L5** contains only the positive differences, which in this case are all four of the differences.

We see that the sum of the ranks of the positive differences is much higher than that of the negative differences. If the medians for each rate were the same, then there would be only a 0.0502 probability of the sum of positive ranks being as high as 10 when expected to be 5 with the four subjects for which there is a difference. The relatively low P-value provides some evidence to reject H_0 and conclude that the median final heart rate is higher for the low rate test.

Exercise 15.19 Below are the readings from Exercise 7.37 of 12 home radon detectors exposed to 105 pCi/l of radon. Apply the Wilcoxon signed rank test to determine if the median reading from all such home radon detectors differs significantly from 105.

91.9	97.8	111.4	122.3	105.4	95.0
103.8	99.6	96.6	119.3	104.8	101.7

Solution. We will test the null hypothesis H_0: median = 105 versus H_a: median $\neq$ 105. First, we enter the given data into list **L1** and then enter 105 twelve times into list **L2**. If H_0 were true, then we would expect the sum of ranked positive differences of **L2 − L1** to be (12 × 13) /4 = 39. But H_a implies that this sum of ranked positive differences will be either much higher than 39 or much lower than 39. To test the hypotheses, we execute the **SIGNRANK** program after entering these data into the lists.

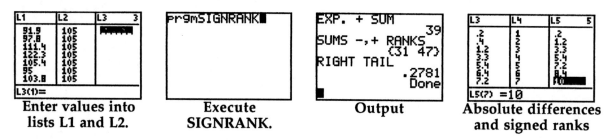

Enter values into lists L1 and L2.	Execute SIGNRANK.	Output	Absolute differences and signed ranks

Lists **L3** and **L4** will show that there were 12 non-zero differences in **L2 − L1**, and list **L5** will show that eight of these were positive differences, meaning that there were eight times in which the home radon detector measured below 105.

The right-tail value is given as 0.2781; thus, the *P*-value for the two-sided alternative is 2 × 0.2781 = 0.5562. If the median home radon measurement were 105, then there would be a 0.5562 probability of the sum of positive ranks being as far away (in either direction) from the expected sum of 39 as the resulting sum of 47 is. Thus, we do not have significant evidence to reject H_0.

15.3 The Kruskal-Wallace Test

Our next program, **KRUSKAL**, is for the Kruskal-Wallace test, which simultaneously compares the distribution of more than two populations. We test the null hypothesis H_0: all populations have same distribution versus the alternative H_a: measurements are systematically higher in some populations. To apply the test, we draw independent SRSs of sizes $n_1, n_2, \ldots, n_I$ from I populations. There are N observations in all. We rank all N observations and let R_i be the sum of the ranks for the ith sample. The Kruskal-Wallace statistic is

$$H = \frac{12}{N(N+1)} \sum_{i=1}^{I} \frac{R_i^2}{n_i} - 3(N+1)$$

When the sample sizes are large and all I populations have the same continuous distribution, then H has an approximate chi-square distribution with $I-1$ degrees of freedom. When H is large, creating a small right-tail *P*-value, then we can reject the hypothesis that all populations have the same distribution.

The KRUSKAL Program

```
PROGRAM:KRUSKAL
:ClrList L₃,L₄,L₅
:dim([B])→L₁
:sum(seq([B](1,J),J,1,L₁(2)))→L
:1→K
:For(J,1,L₁(2))
:For(I,1,[B](1,J))
:[A](I,J)→L₃(K)
:1+K→K
:End:End
:SortA(L₃)
;L₃→L₆:L₃(1)-1→L₃(L+1)
:1→B:1→I
:Lbl 1
:While I<(L)
:If L₃(I)<L₃(I+1)
:Then
:B→L₄(I)
:1+I→I:1+B→B
:Goto 1
:Else
:1→J:B→S
:Lbl 2
:While L₃(I)=L₃(I+J)
:S+B+J→S:1+J→J
:Goto 2
:End
:S/J→T
:For(K,0,J-1)
:T→L₄(I+K)
:End
:I+J→I:B+J→B
:Goto 1
:End:End
```

```
:If I=L
:Then
:L→L₄(I)
:End
:1→K
:Lbl 5
:While K≤L₁(2)
:ClrList L₂
:For(I,1,[B](1,K))
:[A](I,K)→L₂(I)
:SortA(L₂)
:End
:1→I:0→S:0→J
:Lbl 3
:While I≤[B](1,K)
:Lbl 4
:If L₂(I)=L₃(I+J)
:Then
:S+L₄(I+J)→S:1+I→I
:Else
:1+J→J
:End
:Goto 3
:End
:S→L₅(K):1+K→K
:Goto 5
:End
:L₆→L₃:ClrList L₂,L₆
:12/L/(L+1)*sum(seq(L₅(I)²/
   [B](1,I),I,1,L₁(2)))-3(L+1)→W
:1-x²cdf(0,W,L₁(2)-1)→P
:Disp "TEST STAT",W
:Disp "P-VALUE",P
```

Before executing the **KRUSKAL** program, use the **MATRX EDIT** screen to enter the data as columns into matrix [A] and to enter the sample sizes as a row into matrix [B].

Exercise 15.24 Use the Kruskal-Wallace test to see if there are significant differences in the numbers of insects trapped by the board colors.

Board color	Insects trapped					
Lemon yellow	45	59	48	46	38	47
White	21	12	14	17	13	17
Green	37	32	15	25	39	41
Blue	16	11	20	21	14	7

Solution. To execute the **KRUSKAL** program, we must use the **MATRX EDIT** screen to enter the data into matrix [A] and to enter the sample sizes into matrix [B]. First, enter the data into the columns of the 6×4 matrix [A] as you would normally enter data into lists. Next, enter the sample sizes into a 1×4 matrix [B]. Then execute the **KRUSKAL** program.

Enter data as columns in matrix [A].	Enter sample sizes as a row in matrix [B].	KRUSKAL output	Sums of ranks in list L5

After the program completes, view the entries in lists **L3**, **L4**, and **L5**. List **L3** contains the merged, sorted measurements, and list **L4** contains their (averaged) ranks. List **L5** contains the sum of ranks from each type of color. The low *P*-value of 0.00072 gives good evidence to reject the hypothesis that all colors yield the same distribution of insects trapped.

CHAPTER
16

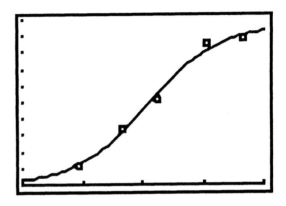

Logistic Regression

16.1 The Logistic Regression Model
16.2 Inference for Logistic Regression

Introduction

In this chapter, we give a brief discussion of two types of logistic regression fits. The first type is a linear fit for the logarithm of the odds ratio of two population proportions. The second type is the general logistic fit for several population proportions.

16.1 The Logistic Regression Model

First, we provide a supplementary program that computes appropriate mathematical odds for a given probability p of an event A. If $p \leq 0.50$, then the odds *against* A are given as the ratio $(1-p):p$. If $p > 0.50$, then the odds *in favor of* A are given as the ratio $p:(1-p)$. The probability can be entered either as a decimal or as a fraction. If p is entered as a decimal, then the odds are computed after rounding p to four decimal places; thus some accuracy may be lost.

The ODDS Program

```
PROGRAM:ODDS
:Menu("ODDS","INPUT DECIMAL",       :Input A
 1,"INPUT FRACTION",2)              :Disp "PROB DENOMINATOR"
:Lbl 1                              :Input B
:Disp "PROBABILITY"                 :A/B→P
:Input P                            :A/gcd(A,B-A)→N
:round(P,4)→P                       :(B-A)/gcd(A,B-A)→D
:10000*P→A                          :Lbl 3
:10000*(1-P)→B                      :If P>.50
:A/gcd(A,B)→N                       :Then
:B/gcd(A,B)→D                       :Disp "ODDS IN FAVOR"
:Goto 3                             :Disp {N,D}
:End                                :Else
:Lbl 2                              :Disp "ODDS AGAINST"
:Disp "PROB NUMERATOR"              :Disp {D,N}
```

Exercise 16.13 In a study of 91 high-tech companies and 109 non-high-tech companies, 73 of the high-tech companies and 75 of the non-high-tech companies offered incentive stock options to key employees.

(a) What proportion of high-tech companies offer stock options to their key employees? What are the odds?
(b) What proportion of non-high-tech companies offer stock options to their key employees? What are the odds?
(c) Find the odds ratio using the odds for the high-tech companies in the numerator.

Solution. (a) The proportion of high-tech companies that offer stock options is simply $73/91 \approx$ 0.8022. To compute the odds, we can use the **ODDS** program. After bringing up the program, enter the numerator value of **73** followed by the denominator value of **91**. The odds in favor are displayed as 73 to 18.

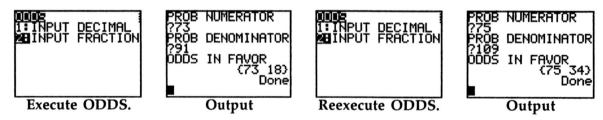

Execute ODDS.   Output   Reexecute ODDS.   Output

(b) For the non-high-tech companies, the proportion is $75/109 \approx 0.688$. Because 34 of these companies do not offer stock options, the odds are $75 : 34$ that such a company does offer stock options to their key employees. This result can be verified with the **ODDS** program.

(c) The odds-in-favor ratio can be computed by simple division of the odds $(73/18) \div (75/34)$. Thus, the odds in favor of a high-tech company offering stock options are about 1.8 times more than the odds for a non-high-tech company.

```
(73/18)/(75/34)
       1.838518519
■
```

The logistic regression model is always based on the odds in *favor* of an event. It provides another method of studying the odds in favor ratio between two populations. As a lead-in, we now provide a program that specifically computes the odds-in-favor ratio.

The ODDS2 Program

PROGRAM:ODDS2 :Disp "1ST PROPORTION" :Input P :Disp "2ND PROPORTION"	:Input R :Disp "ODDS RATIO" :Disp P/(1-P)/(R/(1-R))

Exercise 16.18 (c) In a study on gender bias in textbooks, 48 out of 60 female references were "girl." Also, 52 out of 132 male references were "boy." These two types of references were denoted as juvenile references. Compute the odds ratio for comparing the female juvenile references to the male juvenile references.

Solution. We simply enter the data into the **ODDS2** program that computes the ratio of odds. We see that the odds in favor of a juvenile female response are more than six times the odds in favor of a juvenile male response.

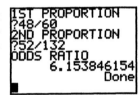

```
1ST PROPORTION
?48/60
2ND PROPORTION
?52/132
ODDS RATIO
       6.153846154
            Done
■
```

Model for Logistic Regression

The logistic regression model is given by the equation

$$\log\left(\frac{p}{1-p}\right) = \beta_0 + \beta_1 x$$

where x is either 1 or 0 to designate the explanatory variable. We now provide a program that computes and displays the regression coefficients for the fit $\log(\text{ODDS}) = b_0 + b_1 x$ as well as the odds ratio e^{b_1}.

The LOG1 Program

```
PROGRAM:LOG1                    :Output(1,6,"A+BX")
:Disp "1ST PROPORTION"          :Output(2,1,"A=")
:Input P                        :Output(2,3,A)
:Disp "2ND PROPORTION"          :Output(3,1,"B=")
:Input R                        :Output(3,3,B)
:ln(R/(1-R))→A                  :Output(5,1,"ODDS RATIO")
:ln(P/(1-P))-A→B                :Output(6,1,e^(B))
:ClrHome
```

Example 16.4 The table below gives data on the numbers of men and women who responded "Yes" to being frequent binge drinkers in a survey of college students. Find the coefficients for the logistic regression model and the odds ratio of men to women.

Population	X	n
Men	1630	7180
Women	1684	9916

Solution. We simply enter the data into the **LOG1** program to obtain $\log(\text{ODDS}) \approx -1.59 + 0.36\,x$. The odds ratio is also displayed.

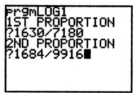

Enter proportions.   Output

16.2 Inference for Logistic Regression

To compute confidence intervals for the slope β_1 and the odds ratio of the logistic regression model, we can use the **ODDSINT** program that follows. To execute the program, separately enter the numerators and denominators of the two sample proportions and the desired level of confidence.

The ODDSINT Program

```
Program:ODDSINT                 :Input L
:Disp "1ST NO. OF YES"          :√(1/M+1/(P-M)+1/N+1/(Q-N))→S
:Input M                        :invNorm((L+1)/2,0,1)→R
:Disp "1ST SAMPLE SIZE"         :ln(M/P/(1-M/P))-ln(N/Q/(1-N/Q))→B
:Input P                        :ClrHome
:Disp "2ND NO. OF YES"          :Disp "SLOPE INTERVAL"
:Input N                        :Disp {round(B-R*S,4),round(B+R*S,4)}
:Disp "2ND SAMPLE SIZE"         :Disp "ODDS RATIO INT."
:Input Q                        :Disp {round(e^(B-R*S),4),
:Disp "CONF. LEVEL"               round(e^(B+R*S), 4)}
```

To test the hypothesis that an odds ratio equals 1, we equivalently can test whether the logistic regression model coefficient b_1 equals 0. To do so, we use the P-value given by

$$P\left(\chi^2(1) \ge (b_1 / SE_{b_1})^2\right)$$

where SE_{b_1} is the standard error of the coefficient b_1. The **ODDSTEST** program that follows computes this P-value upon entering the values of the two proportions under consideration and the value of the standard error SE_{b_1}.

The ODDSTEST Program

`Program:ODDSTEST`	`:ln(R/(1-R))→A`
`:Disp "1ST PROPORTION"`	`:ln(P/(1-P))-A→B`
`:Input P`	`:(B/S)^2→Z`
`:Disp "2ND PROPORTION"`	`:1-x²cdf(0,Z,1)→P`
`:Input R`	`:ClrHome`
`:Disp "ST.ERROR OF B1"`	`:Disp "TEST STAT",Z`
`:Input S`	`:Disp "P-VALUE",P`

Exercise 16.20 In the study on gender bias in textbooks from Exercise 16.18, 48 out of 60 female references were "girl" and 52 out of 132 male references were "boy." The estimated slope is $b_1 = 1.8171$ and its standard error is 0.3686.

(a) Give a 95% confidence interval for the slope.
(b) Calculate the X^2 statistic for testing the null hypothesis that the slope is zero and give the approximate P-value.

Solution. For Part (a) we execute the **ODDSINT** program and for Part (b) we execute the **ODDSTEST** program. We obtain a 95% confidence interval for the slope of {1.0946, 2.5395} that is equivalent to an odds ratio interval of {2.9881, 12.6736}.

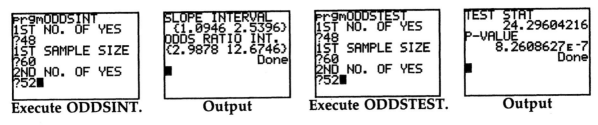

Execute ODDSINT.	Output	Execute ODDSTEST.	Output

Because 0 is not in the slope interval and 1 is not in the odds ratio interval, we have some evidence to reject the null hypothesis that the slope is zero. The X^2 statistic of 24.296 yields a very low P-value of about 8.26×10^{-7} that gives significant evidence to reject the null hypothesis that the slope equals zero.

The Logistic Curve

A general logistic curve is given by the function $p = \dfrac{c}{1 + ae^{-bx}}$. Such a fit can be obtained with the **Logistic** command (item B) from the **STAT CALC** menu. Following is an example to illustrate this fit.

Example An experiment was designed to examine how well an insecticide kills a certain type of insect. Find the logistic regression curve for the proportion of insects killed as a function of the insecticide concentration.

Concentration	Number of insects	Number killed
0.96	50	6
1.33	48	16
1.63	46	24
2.04	49	42
2.32	50	44

Solution. First, enter the data into the **STAT EDIT** screen with the concentrations in list **L1** and the proportions of insects killed in list **L2**. Next, enter the command **Logistic L1,L2,Y1** to compute the regression fit and to store the equation in function **Y1**.

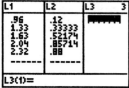
Enter data.

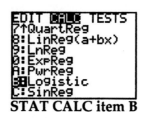
STAT CALC item B

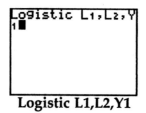
Logistic L1,L2,Y1

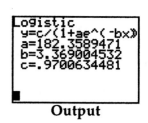
Output

We obtain a logistic regression fit of $y \approx \dfrac{0.97}{1 + 182.36\,e^{-3.369x}}$. If desired, we can make a scatterplot of the data along with the logistics regression curve.

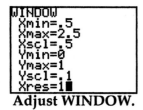
Adjust WINDOW.

Adjust STAT PLOT.

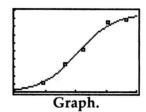
Graph.

CHAPTER
17

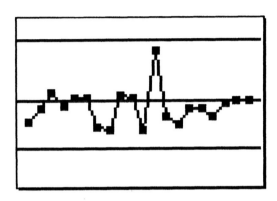

Statistics for Quality: Control and Capability

Introduction

In this chapter, we provide several programs for computing control limits, graphing control charts, and for computing the capability indexes of a process.

17.1 Statistical Process Control

In this section, we provide a program that computes the upper and lower control limits and graphs the control charts for $\bar{x}$ and s.

The CONTRL Program

```
PROGRAM:CONTRL
:Menu("CONTRL","XBAR",1,"S",2,"
 QUIT",3)
:Lbl 1
:Disp "SAMPLE SIZES"
:Input N
:Disp "MEAN"
:Input M
:Disp "STANDARD DEV."
:Input S
:"M"→Y₁
:"M+3*S/√(N)"→Y₂
:"M-3*S/√(N)"→Y₃
:If dim(L₁)>0
:Then
:seq(I,I,1,dim(L₁))→L₃
:0→Xmin
:dim(L₁)+1→Xmax
:Xmax→Xscl
:min(min(L₁),M-4*S/√(N))→Ymin
:max(max(L₁),M+4*S/√(N))→Ymax
:Ymax-Ymin→Yscl
:Plot1(xyLine,L₃,L₁,□)
:End
:ClrHome
:Output(1,4,"XBAR LIMITS")
:Goto 4
:Lbl 2
:Disp "SAMPLE SIZES"
:Input N
:Disp "STANDARD DEV."
:Input S
:If int(N/2)=N/2
:Then
:((N/2-1)!)²*2^(N-2)/(N-2)!*
```

```
√(2/π/(N-1))→C
:Else
:√(2π/(N-1))*(N-1)!/((N-1)/2)!
 /((N-3)/2)!/2^(N-1)→C
:End
:"C*S"→Y₁
:"C*S+3S√(1-C²)"→Y₂
:"max(C*S-3S√(1-C²),0)"→Y₃
:If dim(L₂)>0
:Then
:seq(I,I,1,dim(L₂))→L₃
:0→Xmin
:dim(L₂)+1→Xmax
:Xmax→Xscl
:min(min(L₂),C*S-4S√(1-C²))
 →Ymin
:max(max(L₂),C*S+4S√(1-C²))
 →Ymax
:Ymax-Ymin→Yscl
:Plot1(xyLine,L₃,L₂,□)
:End
:ClrHome
:Output(1,4,"S LIMITS")
:Lbl 4
:PlotsOff
:PlotsOn 1
:AxesOff
:Output(3,1,"UCL")
:Output(3,5,Y₂)
:Output(4,1," CL")
:Output(4,5,Y₁)
:Output(5,1,"LCL")
:Output(5,5,Y₃)
:Lbl 3
:Stop
```

Exercise 17.15 A manufacturer checks the control of a milling process by measuring a sample of five consecutive items during each hour's production. The target width of a slot cut by the milling machine is $\mu = 0.8750$ in. with a target standard deviation of 0.0012 in. What are the centerline and control limits for an s chart? For an $\bar{x}$ chart?

Solution. Bring up the **CONTRL** program and enter either **1** or **2** for the desired variable's control limits. Then enter the sample size of **5** and the target statistics. The centerline and control limits are then displayed.

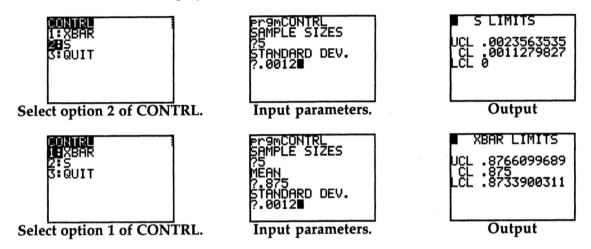

| Select option 2 of CONTRL. | Input parameters. | Output |
| Select option 1 of CONTRL. | Input parameters. | Output |

If we have the values of $\bar{x}$ and s from various samples, then we also can use the program to display the control charts. To do so, always enter the values of $\bar{x}$ into list **L1** and enter the values of s into list **L2**. After executing the **CONTRL** program, press **GRAPH** to see the control chart.

Example 17.4 The following mesh tension data, from Table 17.1 in the textbook, gives the sample mean and sample deviation from 20 different samples of size 4.

Sample mean	Standard deviation	Sample mean	Standard deviation
253.4	21.8	253.2	16.3
285.4	33.0	287.9	79.7
255.3	45.7	319.5	27.1
260.8	34.4	256.8	21.0
272.7	42.5	261.8	33.0
245.2	42.8	271.5	32.7
265.2	17.0	272.9	25.6
265.6	15.0	297.6	36.5
278.5	44.9	315.7	40.7
285.4	42.5	296.9	38.8

The target mean tension is $\mu = 275$ mV with a target standard deviation of 43 mV. Find the centerline and control limits for $\bar{x}$ and for s. Graph the control charts for each.

Solution. Enter the sample means into list **L1** and the standard deviations into list **L2**. Then execute the **CONTRL** program for the desired variable to obtain the control limits, and press **GRAPH** to see the control chart. If desired, press **TRACE** and scroll right to see the individual points.

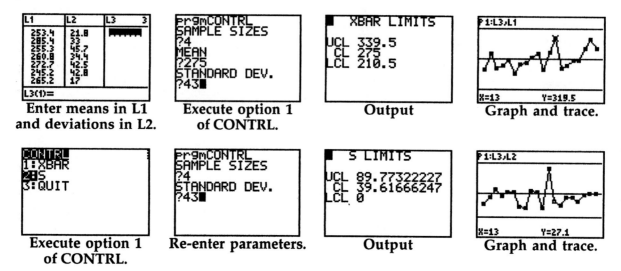

Enter means in L1 and deviations in L2.	Execute option 1 of CONTRL.	Output	Graph and trace.
Execute option 1 of CONTRL.	Re-enter parameters.	Output	Graph and trace.

17.2 Using Control Charts

We now provide a variation of the **CONTRL** program that will compute the upper and lower control limits and graph the control charts for $\bar{x}$ or s based on past data.

Example 17.10 The following data, from Table 17.5 in the textbook, gives the mean and standard deviation of elastomer viscosity from each of 24 samples of size 4.

Sample	$\bar{x}$	s	Sample	$\bar{x}$	s
1	49.750	2.684	13	47.875	1.118
2	49.375	0.895	14	48.250	0.895
3	50.250	0.895	15	47.625	0.671
4	49.875	1.118	16	47.375	0.671
5	47.250	0.671	17	50.250	1.566
6	45.000	2.684	18	47.000	0.895
7	48.375	0.671	19	47.000	0.447
8	48.500	0.447	20	49.625	1.118
9	48.500	0.447	21	49.875	0.447
10	46.250	1.566	22	47.625	1.118
11	49.000	0.895	23	49.750	0.671
12	48.125	0.671	24	48.625	0.895

(a) Find the centerline and control limits for $\bar{x}$ and for s based on this past data. Graph the control charts for each.
(b) Remove the two values of s that are out of control and re-evaluate the control limits for s based on the remaining data.
(c) Remove the corresponding two values of $\bar{x}$ from the samples that were removed in (b) and re-evaluate the control limits for $\bar{x}$ based on the remaining data.

The CONTRL2 Program

```
Program:CONTRL2
:Disp "SAMPLE SIZES"
:Input N
:If int(N/2)=N/2
:Then
:((N/2-1)!)²*2^(N-2)/(N-2)!
 *√(2/π/(N-1))→C
:Else
:√(2π/(N-1))*(N-1)!/((N-1)/2)!
 /((N-3)/2)!/2^(N-1)→C
:End
:Disp "1=XBAR, 2=S"
:Input W
:If W=1
:Then
:2-Var Stats L₁,L₂
:x̄→M
:ȳ/C→S
:"M"→Y₁
:"M+3*S/√(N)"→Y₂
:"M-3*S/√(N)"→Y₃
:seq(I,I,1,dim(L₁))→L₃
:0→Xmin
:dim(L₁)+1→Xmax
:Xmax→Xscl
:min(min(L₁),M-4*S/√(N))→Ymin
:max(max(L₁),M+4*S/√(N))→Ymax
:Ymax-Ymin→Yscl
:Plot1(xyLine,L₃,L₁,□)
:ClrHome
:Output(1,4,"XBAR LIMITS")
```

```
:Goto 4
:Else
:1-Var Stats L₂
:x̄/C→S
:"C*S"→Y₁
:"C*S+3S√(1-C²)"→Y₂
:"max(C*S-3S√(1-C²),0)"→Y₃
:seq(I,I,1,dim(L₂))→L₃
:0→Xmin
:dim(L₂)+1→Xmax
:Xmax→Xscl
:min(min(L₂),C*S-4S√(1-C²))
 →Ymin
:max(max(L₂),C*S+4S√(1-C²))
 →Ymax
:Ymax-Ymin→Yscl
:Plot1(xyLine,L₃,L₂,□)
:ClrHome
:Output(1,4,"S LIMITS")
:End
:Lbl 4
:PlotsOff
:PlotsOn 1
:AxesOff
:Output(3,1,"UCL")
:Output(3,5,Y₂)
:Output(4,1," CL")
:Output(4,5,Y₁)
:Output(5,1,"LCL")
:Output(5,5,Y₃)
```

We note that data sets of *equal* sizes must be entered into lists **L1** and **L2** in order to obtain the control limits for $\bar{x}$. However, only the values of the sample deviations need to be entered into list **L2** in order to compute the control limits for s.

Solution. (a) First, enter the sample means into list **L1** and the sample deviations into list **L2**, then bring up the **CONTRL2** program. Enter 4 for the sample size, then enter **1** when prompted to calculate the control limits for $\bar{x}$ based on this past data. Press **GRAPH** to see the control chart. Then re-execute the program, but enter **2** when prompted to calculate the control limits for s.

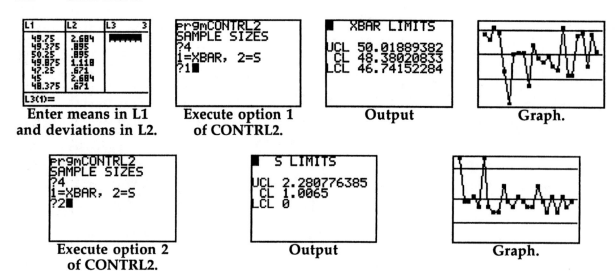

Enter means in L1 and deviations in L2. Execute option 1 of CONTRL2. Output Graph.

Execute option 2 of CONTRL2. Output Graph.

(b) Because the upper control limit for s is 2.28, the values of s from the first and sixth samples are out of control. We now delete these two values, that are both 2.684, from list **L2** and re-execute the **CONTRL2** program.

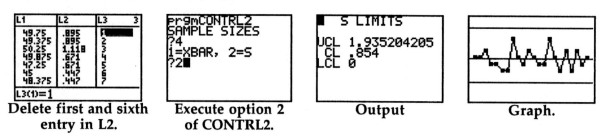

Delete first and sixth entry in L2. Execute option 2 of CONTRL2. Output Graph.

(c) We now delete the values of $\bar{x}$ from the first and sixth samples from list **L1**, then re-execute option 1 of the **CONTRL2** program to find the new control limits for $\bar{x}$.

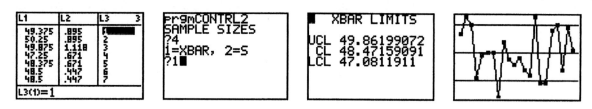

Exercise 17.42 The following data give Joe's weight, measured once each week, for the first 16 weeks after his injury.

Week	1	2	3	4	5	6	7	8
Weight	168.7	167.6	165.8	167.5	165.3	163.4	163.0	165.5

Week	9	10	11	12	13	14	15	16
Weight	162.6	160.8	162.3	162.7	160.9	161.3	162.1	161.0

Joe has a target of $\mu = 162$ pounds for his weight. The short-term variation is estimated to be about $\sigma = 1.3$ pounds. Make a control chart for his measurements using control limits $\mu \pm 2\sigma$.

Solution. Simply make a time plot of the measurements together with graphs of the lines $y = \mu$ and $y = \mu \pm 2\sigma$. To do so, enter the weeks into list **L1** and the weights into list **L2**, then adjust the **STAT PLOT** settings for a scatterplot of **L1** versus **L2**. Next, enter the functions **Y1 = 162+2*1.3**, **Y2 = 162**, and **Y3 = 162−2*1.3** into the **Y=** screen, and adjust the **WINDOW** settings so that the **X** range includes all weeks and the **Y** range includes all weights as well as the upper and lower control limits. Then pres **GRAPH**.

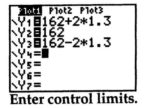
Enter data.

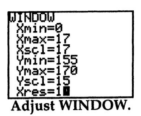
Enter control limits.

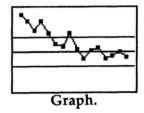
Adjust WINDOW.

Graph.

17.3 Process Capability Indexes

In this section, we provide a program to compute the capability indexes. The program can be used when the parameters are given or when data is given in a list.

The CAPIND Program

```
PROGRAM:CAPIND
:Disp "LSL"                  :ȳ→S
:Input L                     :End
:Disp "USL"                  :Lbl 1
:Input U                     :(U-L)/(6S)→C
:Disp "1=STATS, 2=LIST"      :If M≤L or M≥U
:Input W                     :Then
:If W=1                      :0→D
:Then                        :Else
:Disp "MEAN"                 :min(M-L,U-M)/(3S)→D
:Input M                     :End
:Disp "STANDARD DEV."        :ClrHome
:Input S                     :Output(1,4,"CAP.INDEXES")
:Goto 1                      :Output(3,2,"Cp")
:Else                        :Output(3,5,C)
:2-Var Stats L₁,L₂           :Output(5,1,"Cpk")
:x̄→M                         :Output(5,5,D)
```

Exercise 17.50 Below are data on a hospital's losses for 120 DRG 209 patients collected as 15 monthly samples of eight patients each. The hospital has determined that suitable specification limits for its loss in treating one such patient are LSL = \$4000 and USL = \$8000. Estimate the percent of losses that meet the specifications. Estimate C_p and C_{pk}.

Sample	$\bar{x}$	s	Sample	$\bar{x}$	s
1	6360.6	521.7	9	6479.0	704.7
2	6603.6	817.1	10	6175.1	690.5
3	6319.8	749.1	11	6132.4	1128.6
4	6556.9	736.5	12	6237.9	596.6
5	6653.2	503.7	13	6828.0	879.8
6	6465.8	1034.3	14	5925.5	667.8
7	6609.2	1104.0	15	6838.9	819.5
8	6450.6	1033.0			

Solution. First, enter the sample means into list **L1** and the sample deviations into list **L2**. Then enter the command **2–Var Stats L1,L2** to compute $\bar{x}$ and s. We see that $\bar{x} \approx 6442.43$ and $s \approx 799.127$ (from the $\bar{y}$ output).

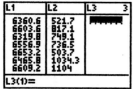

Enter means in L1 and deviations in L2.

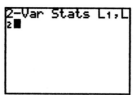

Compute 2-Sample statistics.

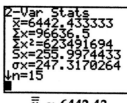

$\bar{\bar{x}} \approx 6442.43$

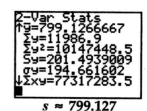

$s \approx 799.127$

Next, use the **normalcdf(** command from the **DISTR** menu to compute $P(4000 \leq X \leq 8000)$ for $X \sim N(6442.43, 799.127)$. We find that about 97.32% of losses meet the specifications. Lastly, bring up the **CAPIND** program. Enter the LSL of 4000 and the USL of 8000, then enter 2 when prompted for **LIST** to output the capability index approximations.

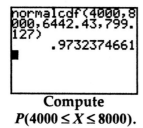

Compute $P(4000 \leq X \leq 8000)$.

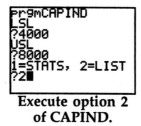

Execute option 2 of CAPIND.

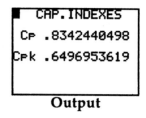

Output

Exercise 17.51 The dimension of the opening of a clip has specifications 15 ± 0.5 millimeters. The production of the clip is monitored by $\bar{x}$ and s charts based on samples of five consecutive clips each hour. The 200 individual measurements from the past week's 40 samples have $\bar{x} = 14.99$ mm and $s = 0.2239$ mm.

(a) What percent of clip openings will meet specifications if the process remains in its current state? (b) Estimate the capability index C_{pk}.

Solution. (a) We must compute $P(15-0.5 \leq X \leq 15+0.5)$ for $X \sim$ $N(14.99, 0.2239)$. Using the **normalcdf(** command, we find that about 97.43% of clip openings will meet specifications if production remains in its current state.

(b) To estimate C_{pk}, we shall use the **CAPIND** program. Bring up the program and enter the LSL of 14.5 and the USL of 15.5, then enter **1** when prompted for **STATS**. Enter the mean and standard deviation to receive the output. We find that $C_{pk} \approx 0.7295$.

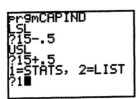  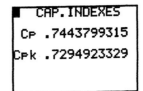

17.4   Control Charts for Sample Proportions

We conclude this chapter with a program to compute the control limits for sample proportions. The program can be used with summary statistics or when a list of data is given.

The CONTRLP Program

```
PROGRAM:CONTRLP
:Menu("CONTRLP","STATS",1,"LIST",2,"
 QUIT",3)
:Lbl 1
:Disp "TOTAL SUCCESSES"
:Input T
:Disp "NO. OF STAGES"
:Input M
:Disp "NO. PER STAGE"
:Input N
:T/(M*N)→P
:Goto 4
:Lbl 2
:1-Var Stats L₂
:x̄→N
:sum(seq(L₁(I),I,1,dim(L₁)))/sum(
 seq(L₂(I),I,1,dim(L₂)))→P
:L₁/L₂→L₃
:seq(I,I,1,dim(L₁))→L₄
:Lbl 4
:"P"→Y₁
:"min(P+3*√(P(1-P)/N),1)"→Y₂
:"max(P-3*√(P(1-P)/N),0)"→Y₃
:0→Xmin

:dim(L₁)+1→Xmax
:Xmax→Xscl
:min(Y₃,min(L₃))-.01→Ymin
:max(Y₂,max(L₃))+.01→Ymax
:1→Yscl
:Plot1(xyLine,L₄,L₃,□)
:PlotsOff
:PlotsOn 1
:AxesOff
:ClrHome
:Output(1,5,"P LIMITS")
:Output(3,1,"PBAR")
:Output(3,6,P)
:Output(4,1,"NBAR")
:Output(4,6,N)
:Output(6,1,"UCL")
:Output(6,5,Y₂)
:Output(7,1," CL")
:Output(7,5,Y₁)
:Output(8,1,"LCL")
:Output(8,5,Y₃)
:Lbl 3
:Stop
```

Exercise 17.59 Over the last ten months, an average of 2875 invoices per month have been received with only a total of 960 remaining unpaid after 30 days. Find $\bar{p}$. Give the centerline and control limits for a p chart.

Solution. Bring up the **CONTRLP** program and press **1** for **STATS**. Enter the values of **960** for total "successes," **10** for the number of "stages," and **2875** for the number per stage. We find that $\bar{p} \approx 0.0334$ with an LCL of 0.02334 and a UCL of 0.04344.

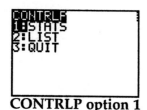
CONTRLP option 1

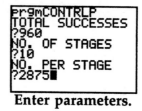
Enter parameters.

P LIMITS
PBAR .0333913043
NBAR 2875
UCL .0434431167
CL .0333913043
LCL .023339492
Output

Exercise 17.63 Here are data on the total number of absentees among eighth graders at an urban school district.

Month	Sep.	Oct.	Nov.	Dec.	Jan.	Feb.	Mar.	Apr.	May	Jun.
Students	911	947	939	942	918	920	931	925	902	883
Absent	291	349	364	335	301	322	344	324	303	344

(a) Find $\bar{p}$ and $\bar{n}$. (b) Make a p chart using control limits based on $\bar{n}$ students each month.

Solution. (a) First, enter the numbers of absentees ("successes") into list **L1** and the numbers of students into list **L2**. Next, execute option 2 of the **CONTRLP** program. We find that $\bar{p} \approx$ 0.3555 and $\bar{n} = 921.8$. (b) After executing the program, the individual monthly proportions are stored in list **L3**. Press **GRAPH** to see the p chart that has an LCL of 0.3082 and a UCL of 0.4028.

Enter data into
lists L1 and L2.

CONTRLP
1:STATS
2:LIST
3:QUIT
Execute CONTRLP
option 2.

P LIMITS
PBAR .3555001085
NBAR 921.8
UCL .402797174
CL .3555001085
LCL .308203043
Output

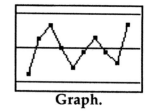
Graph.

Index of Programs
Programs can be downloaded at
http://www.wku.edu/~david.neal/ips5e/

ANOVA1 (page 93) displays the overall sample mean, the pooled sample deviation, the mean square error for groups MSG, the mean square for error MSE, the R^2 coefficient, the F-statistic, and the P-value of the ANOVA test for equality of means when data is entered as summary statistics. Before executing the program, enter the sample sizes into list **L1**, the sample means into list **L2**, and the sample deviations into list **L3**.

ANOVA2 (page 99) displays the P-values for two-way analysis of variance. For one observation per cell, enter the data into matrix **[A]** before executing the program. For c observations per cell, enter the means into matrix **[A]** and the standard deviations into matrix **[B]**. The program also stores the marginal means for the rows and columns into lists **L2** and **L4**. The overall mean is stored as the first value in list **L5**. The remainder of **L5** are the values SSA, SSB, and SSE.

ANPOWER (page 96) computes the Winer approximation of the power of the ANOVA test under the alternative H_a that the true population means are μ_1, μ_2, . . ., μ_I. Before executing the program, enter the successive sample sizes n_1, . . . , n_I into list **L1** and the alternative population means into list **L2**. After the level of significance and the guessed standard deviation are entered in the running of the program, the approximate power is displayed along with the values of F^*, DFG, DFE, and the noncentrality parameter λ.

BAYES (page 32) computes the total probability $P(C)$ and conditional probabilities associated with Bayes' rule. Before executing the program, enter values for $P(A_i)$ into list **L1** and the conditional probabilities $P(C \mid A_i)$ into list **L2**. The program displays $P(C)$, stores $P(C \cap A_i)$ in list **L3**, stores $P(A_i \mid C)$ in list **L4**, stores $P(A_i \mid C')$ in list **L5**, and stores $P(C \mid A_i')$ in list **L6**.

BOOT (page 104) performs resampling on a random sample that has been entered into list **L1**. If a bootstrap confidence interval for the statistic is desired, enter **1** for **CONF. INTERVAL?** when prompted; otherwise, enter **0**. The program takes resamples from the entered random sample and enters their means into list **L2**. The mean of all resamples, the bootstrap standard error, and the confidence interval (if specified) are displayed.

BOOTCORR (page 111) performs the bootstrap procedure on the correlation coefficient or the regression slope for paired sample data that has been entered into lists **L1** and **L2**. When prompted, enter **1** if you want to bootstrap the correlation coefficient or enter **2** if you want to bootstrap the regression slope. The resampled statistics are stored in list **L3**. The statistic of the original sample data is displayed along with the bootstrap standard error and the confidence interval.

BOOTPAIR (page 108) computes a bootstrap t confidence interval for the difference in means based on random samples that have been entered into lists **L1** and **L2**. The resampled differences in mean are stored in list **L3**. The difference of the original sample averages is displayed along with the bootstrap standard error and the confidence interval.

BOOTTEST (page 114) performs a permutation test for the difference in means. Before executing, enter data from the first population into list **L1** and enter data from the second population into list **L2**. When prompted, enter **1, 2,** or **3** to designate the desired alternative. The resampled differences in permuted mean are ordered and then stored in list **L3**. The program displays the difference of the original sample means and the P-value.

BOOTTRIM (page 107) computes a bootstrap t confidence interval for a trimmed mean on a random sample that has been entered into list **L1**. When prompted, enter the desired number of resamples, the decimal amount to be trimmed at each end, and the desired confidence level. The program takes resamples from the entered random sample and enters their trimmed means into list **L2**. The trimmed mean of the original sample, the trimmed bootstrap standard error, and the confidence interval are displayed.

BTPRTEST (page 116) performs a permutation test for either the difference in paired means or for the correlation. Before executing, enter the data set into lists **L1** and **L2**. When prompted, enter **1** or **2** to designate the desired test, then enter **1, 2,** or **3** to designate the desired alternative. The resampled permuted pair differences in mean (or correlations) are ordered and stored in list **L3**. The statistic from the original paired sample is displayed along with the P-value.

CAPIND (page 141) computes capability indexes based on given parameters or on data that has been entered into lists **L1** and **L2**.

CONTRAST (page 95) computes the P-value for a significance test and a confidence interval for mean population contrasts. Before executing the program, enter the sample sizes into list **L1**, the sample means into list **L2**, the sample deviations into list **L3**, and the contrast equation coefficients into list **L4**. When prompted during the program, enter either **1** or **2** for a one-sided or two-sided alternative.

CONTRL (page 136) computes the upper and lower control limits and graphs the control charts for $\bar{x}$ and s.

CONTRL2 (page 139) computes the upper and lower control limits and graphs the control charts for $\bar{x}$ or s based on past data. Data sets of equal sizes must be entered into lists **L1** and **L2** in order to obtain the control limits for $\bar{x}$. But only the values of the sample deviations need to be entered into list **L2** in order to compute the control limits for s.

CONTRLP (page 143) computes the control limits for sample proportions given either summary statistics or data entered into lists **L1** and **L2**.

DISTSAMP (page 30) draws a random sample from a discrete distribution that has been entered into lists **L1** and **L2**.

FITTEST (page 75) performs a goodness of fit test for a specified discrete distribution. Before executing, enter the specified proportions into list **L1** and enter the observed cell counts into list **L2**. The expected cell counts are computed and stored in list **L3**, and the individual contributions to the chi-square test statistic are stored in list **L4**. The program displays the test statistic and the P-value.

KRUSKAL (page 127) performs the Kruskal-Wallace test. Before executing, enter the data into the columns of matrix **[A]** and the sample sizes into a row matrix **[B]**. The program displays the test statistic and P-value. List **L3** will contain the merged, sorted measurements, **L4** will contain their (averaged) ranks, and **L5** will contain the sum of ranks from each population.

LOG1 (page 132) computes and displays the coefficients of the linear regression model for the log of odds ratio $\log(p/(1-p)) = \beta_0 + \beta_1 x$. Also displays the odds ratio.

MULTREG (page 86) computes the regression coefficients and an ANOVA table for the multiple linear regression model $\mu_y = \beta_0 + \beta_1 x_1 + \ldots + \beta_p x_p$. The squared correlation coefficient, F-statistic, P-value, and the standard deviation are also displayed. Before executing the program, enter sample data as columns into matrix **[A]** with the last column used for the dependent variable.

ODDS (page 130) computes the appropriate mathematical odds for a given probability p of an event A. If $p \le 0.50$, then the odds against A are given as the ratio $(1-p) : p$. If $p > 0.50$, then the odds in favor of A are given as the ratio $p : (1-p)$.

ODDS2 (page 131) computes the odds-in-favor ratio between two proportions.

ODDSINT (page 132) computes a confidence interval for the slope β_1 of the logistic regression model and the odds ratio.

ODDSTEST (page 133) computes the test statistic and P-value for the hypothesis test that an odds ratio equals 1.

POWER2T (page 61) computes a standard normal approximation of the power of the pooled two sample t test upon entering values for the alternative mean difference, the two sample sizes, the level of significance, and the common standard deviation.

PSAMPSZE (page 65) computes the required sample size that would give a maximum desired margin of error m for a proportion confidence interval.

RANDOM (page 21) randomly chooses a subset of specified size from the set $\{1, 2, \ldots, n\}$ and stores the values in list **L1**.

RANKSUM (page 121) performs the Wilcoxon rank sum test on data from two populations. Before executing, enter the data into lists **L1** and **L2**. The program displays the expected sum of ranks from list **L1**, the sums of the ranks from each list, and the smallest tail value created by the test statistic which is the sum of the ranks from **L1**. List **L3** then contains the merged, sorted measurements, and **L4** contains their (averaged) ranks.

REG1 (page 79) computes confidence intervals for the regression slope and intercept. Before executing the program, data must be entered into lists **L1** and **L2**.

REG2 (page 80) computes a confidence interval for a mean response or a prediction interval for an estimated response. Before executing the program, enter paired data into lists **L1** and **L2**.

REG3 (page 82) computes the ANOVA table for linear regression and displays the associated F-statistic and P-value. Before executing the program, data must be entered into lists **L1** and **L2**. The ANOVA table is stored into lists **L4**, **L5**, and **L6**.

SIGNRANK (page 124) performs the Wilcoxon signed rank test on data sets of size n from two populations. Before executing, enter the data into lists **L1** and **L2**. The program sorts the absolute value of the differences **L2 − L1** into list **L3**, but disregards any zero differences. The (averaged) rank of each non-zero difference is stored in list **L4**. The sums of the ranks of the positive differences and of the negative differences are displayed. The program also displays the smallest tail value created by the test statistic which is the sum of the ranks of the positive differences.

TPOWER (page 54) computes the power against an alternative for hypothesis tests about the mean when using a known standard deviation and critical values $t*$.

TSCORE (page 50) finds the critical value $t*$ of a t distribution upon specifying the degrees of freedom and confidence level.

TWOTCI (page 55) computes a confidence interval for the difference of means of normally distributed populations when the critical value $t*$ is obtained from the t distribution having degrees of freedom that is the smaller of $n_1 - 1$ and $n_2 - 1$.

TWOTTEST (page 56) performs hypothesis tests for the difference of means of normally distributed populations when the critical value $t*$ is obtained from the t distribution having degrees of freedom that is the smaller of $n_1 - 1$ and $n_2 - 1$.

TWOWAY (page 70) converts a two-way table of raw data into three different proportion tables. Before executing the program, enter the raw data into matrix [A]. The joint distribution is then stored in matrix [B], the conditional distribution on the column variable is stored in matrix [C], and the conditional distribution on the row variableis stored in matrix [D].

ZPOWER (page 48) computes the power against an alternative for hypothesis tests about the mean when using a known standard deviation and normal distribution z-scores.

ZSAMPSZE (page 43) computes the sample size needed to obtain a desired maximum margin of error with a specified level of confidence when finding a confidence interval for the mean using a known standard deviation and normal distribution z-scores.